621.3
Wo

VGM Opportunities Series

OPPORTUNITIES IN
ELECTRICAL TRADES

Robert Wood

Foreword by
Charles H. Pillard
President Emeritus
International Brotherhood of Electrical Workers

VGM Career Horizons
a division of *NTC Publishing Group*
Lincolnwood, Illinois USA

Cover Photo Credits:
All photos are courtesy of the International
Brotherhood of Electrical Workers.

Library of Congress Cataloging-in-Publication Data

Wood, Robert B., 1934-
  Opportunities in electrical trades / Robert Wood ; foreword by
Charles H. Pillard. — Rev. ed.
      p.   cm. — (VGM opportunities series)
  Includes bibliographical references.
  ISBN 0-8442-8587-0 : $12.95 — ISBN 0-8442-8588-9 (soft) : $9.95
  1. Electric engineering—Vocational guidance.   I. Title.
II. Series.
TK159.W66   1990
621. 3'023'73—dc20   90-33446
                                                                  CIP

Published by VGM Career Horizons, a division of NTC Publishing Group.
© 1990 by NTC Publishing Group, 4255 West Touhy Avenue,
Lincolnwood (Chicago), Illinois 60646-1975 U.S.A.
All rights reserved. No part of this book may be reproduced, stored
in a retrieval system, or transmitted in any form or by any means,
electronic, mechanical, photocopying, recording or otherwise, without
the prior permission of NTC Publishing Group.
Manufactured in the United States of America.

0 1 2 3 4 5 6 7 8 9 VP 9 8 7 6 5 4 3 2 1

ABOUT THE AUTHOR

Robert Wood has been a member of the International Brotherhood of Electrical Workers for over 35 years. His electrical background encompasses experience in the following areas: electrical construction, maintenance, electric motor repair and armature rewinding, marine and ship repair, and industrial electronics.

Mr. Wood completed his apprenticeship training by working on the job during the day and attending related classes at night. Later, he returned to night school and earned certification in the field of electronics.

Upon completion of his apprenticeship, Mr. Wood worked for a number of years as a journeyman electrician, and on several occasions was employed in supervisory positions. After acquiring several years of experience at the journeyman level (and passing the required examination), he received a Class A Electrician's License for the City of New Orleans.

Mr. Wood's background also includes seven years of teaching experience. He was a member of the faculty of Delgado College of New Orleans, where he taught at both the trade school and junior college levels. During his teaching tenure, he taught skill-improvement courses at night to journeyman electricians and developed a laboratory manual for use in an electrical machinery course.

A member of the United States Coast Guard Reserve, Mr. Wood has often served as instructor in marine and shipboard electrical systems. He holds a commission in the Coast Guard Reserve, where he has served as educational and training officer.

In addition to the diplomas he received for his studies of electricity and electronics, Mr. Wood was awarded an associate degree in electrical engineering from Delgado College and a bachelor of science degree in industrial education from Northwestern State College of Louisiana. He has also studied at the University of Virginia.

Mr. Wood now serves as a member of the Labor Research Advisory Council of the United States Department of Labor, Bureau of Labor Statistics, the Labor Sector Advisory Committee for Trade Negotiations, as well as the United Way's Environmental Scan Committee.

ACKNOWLEDGMENTS

I wish to thank my wife, Joyce, for her perseverance in typing the manuscript of this book, and all my former students who afforded me the opportunity of sharing in a mutual learning process—education. Especially, I would like to tender my appreciation to Kenneth Edwards, Director of Skill Improvement for the International Brotherhood of Electrical Workers, for his valuable assistance.

I also want to express my gratitude to the following organizations which so willingly cooperated in supplying information that added to the value of the publication: the International Brotherhood of Electrical Workers, the National Electrical Contractors Association, the National Electric Sign Association, the Electric Apparatus Service Association, the Edison Electric Institute, the International Association of Electrical Inspectors, the National Fire Protection Association, and National Electrical Manufacturers Association.

PREFACE

Thousands of young people will be seeking employment in the electrical industry during the next decade. For most of them, the decision to do so will mark a pivotal point in their lives. Consequently, current information is critical in assisting them in making the proper choice.

Occupational selection should be based on many factors. The individual's ability and interests should be weighed heavily. Job potential, the nature of the work, opportunities for advancement, employment security, earnings, and general working conditions are other important elements to consider. Most importantly, rewarding and productive careers depend on being given opportunities and challenges in line with individual talents.

Prudent career planning depends not only on individuals' assessment of their abilities and desires, but also on the advice of professionals and other experienced people. The purpose of this book is twofold—to serve prospective careerists in their decision-making process and to serve as a source of information to persons having a role in career development—counselors, teachers, parents, and counselor-educators.

The book includes topics of interest to individuals contemplating careers in the electrical trades. Beginning with an overview of the electrical industry, the book then covers guidelines for occupational planning and selection, apprising the student of techniques utilized in determining occupational compatibility. Chapter 3 relates to education and training. Selected occupations in the field, opportunities for advancement, and specialization are treated in Chapters 4 through 6. For the most part, Chapter 7 is statistical in nature, covering such areas of interest as wages and fringe benefits, hours of work, and future work force requirements. The final chapter serves to introduce readers to various organizations in the field.

FOREWORD

Descriptive words such as vast, intricate, dynamic, and innovative are sometimes used to characterize the structure of the electrical industry. However, these words alone do not begin to convey how important this industry has become to our daily lives and to the functions of our nation. Our homes, offices, schools, hospitals, factories, commercial and industrial buildings, streets, highways, bridges, transportation systems, and space programs all utilize electrical energy in one or more forms. We use electricity for illumination and heat and as a source of power to operate electrical machinery and equipment. The various applications stated here enumerate only a few of the many places where we find electrical energy in use in today's technologically advanced world.

The electrical industry is made up of a number of major branches, and each branch performs a significant role in making electrical energy available for our needs. Because electricity can be effectively transferred from one place to another and because it uniquely qualifies as the source of energy for many of our domestic, commercial, and industrial processes and uses, the electrical industry has experienced continuous growth since its commercial value was first discovered.

Foreword ix

The decade of the 1990s holds many challenges for our nation and our economic system. As our country strives for greater real economic growth, the electrical industry will play a vital role. For example, many energy-related benefits will be derived from the construction of energy efficient dwellings with sophisticated control systems, intelligent commercial buildings, industrial buildings designed with the help of computers, and systems incorporating the latest microelectronic technology.

Electrified rapid transit systems will become increasingly important to metropolitan areas, as our nation's cities deal with the escalating problem of traffic congestion and gridlock. By reducing the need for private automobile transportation, these efficient transportation systems accomplish another important function—they reduce the demand for petroleum products. Since the electrical industry will have a paramount role in meeting our energy needs, many opportunities will be available for those interested in this industry.

Mr. Wood has a notable background in the electrical industry. He began his electrical career by completing the national apprenticeship training program for inside construction journeymen. His subsequent work experience includes employment in the following branches of the electrical industry: construction, maintenance, electric motor repair, and marine work. Also, he has taught in both apprentice and journeyman training programs. He has been a member of the International Brotherhood of Electrical Workers (IBEW) for over 35 years. For the past 22 years he has served on the staff of the International Office of the IBEW in the Research and Education Department. Since 1972, he has served as director of that department. Hence, it is my opinion that the reader will benefit greatly from the author's experiences which have been interwoven into the fabric of this text.

Each branch of the electrical industry consists of certain occupations that have evolved due to the nature of the work performed in that particular sector of the industry. In this book, Mr. Wood discusses a number of key occupations found in the electrical industry. He has oriented the material to guide the reader to a better understanding of what is involved in career selection, preparation, and development. Consequently, those who are interested in a career in the electrical industry will find the material covered both interesting and informative.

Charles H. Pillard
President Emeritus
International Brotherhood
of Electrical Workers

CONTENTS

About the Author iii

Acknowledgments v

Preface vi

Foreword viii

1. **View of the Field** 1

    An overview of the electrical industry. Historical development. Current trends and future projections. Earnings and general working conditions. Desirable attributes and assets.

2. **Occupational Planning and Selection** 17

    Education and training. Employment services. Assessing your own abilities. Seeking the services of a guidance counselor.

xi

xii *Opportunities in Electrical Trades*

3. **Developing Your Abilities Through Education and Training** 23
   Prerequisite courses. Industrial arts. Vocational and industrial education. Sample apprenticeship schedules. Apprenticeship training.

4. **Occupations in the Electrical Field** 43
   Construction electrician. Maintenance electrician. The electric motor shop journeyman. Marine electrician. Utility and power plant occupations. Electric sign serviceperson.

5. **Opportunities for Advancement** 65
   Apprenticeship programs. On-the-job training. Evening extension classes. Correspondence courses. Skill improvement programs.

6. **Specializing in Related Fields** 75
   Control specialist. Electrical Serviceperson. Electronics journeyman. Instrumentation technician. Communications mechanic. Sales or service representative. Supervisor. Estimator. Electrical inspector. Electrical contractor. Engineer. Teacher.

7. **Employment and Earnings** 93
   Wages. Fringe benefits. Hours of work. Location of employment. Employment security. Job openings. Tomorrow's jobs.

8. **Organizations in the Field** 107
   Organized labor. International Brotherhood of Electrical Workers. National Electrical Contractors Association. National Electric Sign Association. Electrical Apparatus Service Association. Edison Electric In-

stitute. International Association of Electrical Inspectors. National Fire Protection Association. National Electrical Manufacturers Association. Regulatory and licensing authorities.

Appendix A. Sources of Additional Information **123**

Appendix B. Bureau of Apprenticeship and Training . . **125**

Appendix C. Helpful Publications **135**

Appendix D. Glossary **137**

Appendix E. State Apprenticeship Agencies **139**

CHAPTER 1

VIEW OF THE FIELD

A vocation in the electrical trades can be both demanding and rewarding, primarily because of the range of skills and disciplines which can be mastered in the space of a lifetime in the field.

Selecting a vocation is one of the most important decisions people face as they prepare to enter the work force. Too often people attempt to choose a career without becoming acquainted with the basic structure of their chosen field. Often, little concern is given to whether or not the person has a natural inclination for the occupation selected.

Occupations which appear exciting or glamorous may lead individuals into emotion-charged but unsound career decisions. Some young people select a trade simply because a friend happens to be pursuing it or because a particular occupation is currently popular. A better understanding of the complexity of our industrial society would help these persons make more intelligent decisions as to the trades for which they are best suited and in which they are most likely to be successful.

Our nation's labor force is made up of approximately 116 million workers engaged in a multitude of occupations. Each of these occupations contributes to the total effectiveness of a balanced labor force. And each shares a common interest in the industrial growth of our country.

The vast and complex economy of the United States offers numerous career choices. Thousands of different types of jobs are available, with a wide range of employers. Moreover, the United States Department of Labor, Bureau of Labor Statistics projects that the total labor force will reach about 133 million by the year 2000. This means that the present labor force will be expanded by about 15 percent.

In addition to asking about employment opportunities, young people preparing to enter the work force are seeking other important information. Some typical questions raised by prospective job seekers are: What are the current trends and future projections for the particular field of interest? How do the earnings and working conditions compare with those in other fields? What are the educational requirements for entering the field? What type of training is required in addition to on-the-job training? Are the opportunities for advancement somewhat limited, or are they adequate? What are the desirable attributes for successful career development? Of course, many more questions will undoubtedly arise as one endeavors to select a career that will result in satisfaction and success.

It is the author's aim in the pages that follow to provide meaningful responses to these questions while furnishing information that is relevant to employment opportunities within the electrical industry.

AN OVERVIEW OF THE ELECTRICAL INDUSTRY

The electrical industry is a single, but important, segment of the contract construction industry, which includes establishments engaged primarily in the contract construction of new work, additions, alterations, and repairs. In its Standard Industrial Classification (SIC) Manual, the Office of Management and Budget (OMB) lists three broad types of contract construction establishments: general building contractors engaged primarily in the construction of residential, farm, industrial, commercial, and public or other buildings (SIC 15); general contractors engaged in such heavy construction as highways and streets, bridges, sewers, railroads, irrigation projects, flood control projects, and marine construction (SIC 16); and special trade contractors who undertake such specialized activities as plumbing, painting, plastering, carpentering, and electrical work (SIC 17).

Special trade contractors may work on subcontract from a general contractor or directly for the owner. For the most part, they perform their work at the construction site, although they also may have shops where work incidental to the job site is performed.

Electrical work, classified by the SIC Manual as group No. 1731, is composed of special trade contractors primarily engaged in electrical work at the site, including the following:

Burglar alarm installation
Cable splicing (electrical)
Communication equipment installation
Fire alarm installation
Heating equipment installation (electrical)
Intercommunication equipment installation
Sound equipment installation
Telephone installation

Utility Branch

From a functional or operational point of view, one might say that the electrical industry begins with its utility branch. The initial generation of electricity takes place in utility plants, providing us with the energy that heats and lights our homes, powers our appliances, powers our factories and industrial plants, motorizes giant machinery, fires missiles from their launching pads, and allows us to perform numerous tasks with very little physical effort.

The generation and distribution of electrical power is the primary function of the utility branch, which is constantly growing to meet society's demands for electrical energy. As our population grows, the demand for housing increases, thereby creating a need for more electricity. Since the latter part of the 1970s, the incremental use of electricity has slowed from its previous growth rate of doubling every eight years. Nonetheless, as our nation grows and our economy expands, the utility branch of the electrical industry will obviously need to grow in a commensurate fashion to meet society's needs. Moreover, the share of electricity in the nation's energy profile will continue to expand, rising from around 19 percent in 1987 to more than 23 percent by the year 2015.

Construction

The construction branch of the electrical industry covers the installation of electrical circuits, devices, appliances, and equipment, and is responsible for the final delivery of electrical power for consumer use. As the nation's generating capacity increases, more construction work must take place to allow for this growth. The annual dollar volume of electrical construction currently exceeds $34 billion; it is projected to continue to rise through the forseeable future.

The electrical construction industry is usually separated into two distinct categories: inside construction and outside construction. The inside construction branch can be further subdivided into three widely recognized segments: residential, commercial, and industrial.

Residential jobs cover electrical work installed in houses and related types of buildings. *Commercial construction* is electrical work performed on such commercial establishments as supermarkets, shopping centers, drugstores, service stations, and garages. *Industrial work*, often loosely termed "large work," involves electrical work in the construction of petroleum, chemical, textile, rubber, and metallurgical plants.

According to Census Bureau figures, there were 49,436 inside electrical contractors in 1982 employing approximately 500,000 construction workers. Each type of construction employs its own specialized installation techniques and materials. Further, unit labor requirements are proportionate to the size and type of job; residential installations usually do not require more than two workers, while the average commercial job generally requires five to eight workers, and large industrial jobs frequently utilize more than fifty journeyman electricians.

Outside construction pertains to the assembly, installation, operation, service, and repair of all electrical equipment, wiring, raceways, and related devices outside of closed structures.

Maintenance

Electrical maintenance workers service and repair the electrical machines, equipment, and circuits that keep our nation's power plants, factories, hospitals, schools, apartments, and office buildings operating effectively. This sector of the industry is charged with the responsibility of ensuring safe, steady, and reliable electrical service. To accomplish this task, they routinely perform preventative maintenance on electrical equipment and systems.

Specialties

The electrical industry also encompasses a number of sectors that are primarily engaged in performing specialized functions. The electric motor shop industry, marine and shipboard electrical work, and the installation, repair, and service of electrical signs are typical examples of such specialties. Cable television (CATV) is another example. This branch of the industry is presently experiencing rapid growth. Cable television makes it possible for people living in rural and distant areas to enjoy quality reception of TV programs. It is also a medium used to channel specific entertainment programs, such as movies or sports events, to subscribers for a monthly fee.

Communications

The communications field forms an integral part of the electrical industry. It is essentially concerned with the installation, operation, repair, and service of electronic transmitting, receiving, and recording equipment. Workers in this field set up intricate outside and inside communication systems. They install various types of cables capable of carrying hundreds of telephone messages simultaneously—thus providing our country with one of the finest communication networks in the world.

Automation and modern technology have brought a number of innovations to this field. Much of the manual equipment and many of the physical operations once required to perform a specific function have been completely replaced by automated equipment capable of functioning instantaneously and with a high degree of precision.

Fast moving research and development is bringing fiber-optic technology to the forefront in this industry. Fiber-optic systems involve the transmission of light signals over glass fibers. Just as electrical signals are transmitted over copper wire, now light signals can be transmitted over glass filaments. Fiber-optic cable is flexible and lightweight. Cables with 144 optical fibers and the capability of carrying more than 50,000 telephone conversations simultaneously are no bigger around than your finger. Although this technology is relatively new, telephone companies are making considerable use of it.

Manufacturing

Finally, we reach that sector of the industry that makes it possible for all the other segments to exist—the manufacturing sector. Without it, there would be no conventional means for generating electricity, no medium for distributing it, and no devices and appliances to create a need for it. Workers employed in manufacturing plants are responsible for the production of an infinite variety of electrical and electronic items that range from simple household fuses to complex and sophisticated electronic computers.

The combination of these branches makes for an intricate, dynamic industry that encompasses a multitude of occupations. Basically, however, our object will be to explore those branches that provide employment opportunities for skilled electrical workers.

HISTORICAL DEVELOPMENT

During the latter part of the 19th century, many scientists and experimenters attempted to produce electrical lighting on a practical basis, and records show the occasional use of electric lamps during that period. However, the success of these early products was usually short-lived. It was not until 1879, when Thomas

Edison developed his incandescent lamp with a carbonized filament enclosed in an evacuated glass envelope, that the electric lamp began to produce fruitful results. Then, with the advent of a successful lamp and the development of the dynamo (now known as the electric generator), the need emerged for a system capable of delivering electrical power to densely populated communities.

In the 1880s, commercial supplies of electricity began to appear. Although the first generating stations were rather small in capacity, they helped set the stage for future electrification of our cities. In fact, on September 4, 1882, Thomas Edison announced that he was ready to test a new lighting system which occupied one square mile of New York City. Edison's venture was a success, and so created the need for a new type of worker.

The duties of early electrical workers pertained primarily to the erection of distribution systems. These systems included the conductors, commonly called lines, and the poles used to support them. It was from this type of work that the "lineperson" job classification originated. The lineperson's work was exciting and interesting, but it was also very dangerous. The fatality rate was high, and insurance was virtually impossible to obtain. The work was hard, the hours were long, and the pay was low. Fortunately, there were enough dedicated, courageous people to lay the foundation upon which today's electrical industry is built.

As electrical power became available to more communities, the need for a larger, more versatile work force grew. While the lineperson continued to install and maintain the transmission and distribution systems, a demand developed for electrical workers skilled in wiring residences, buildings, and other commercial establishments. These workers had to be knowledgeable in the art of installing electrical circuits and equipment in a safe, professional manner. Although the tools of the trade were rather simple in those days, workers still had to be trained in the art of using them skillfully. These workers were essentially concerned with wiring buildings, and since a large part of their work was performed inside these buildings, they became known as *inside journeymen*. Today, these highly skilled workers comprise one of the leading groups in the building trades division of the construction industry.

As technology advanced, industry began to appreciate the advantages of electricity for power. Soon, more and more manufacturing firms began to electrify their operations, reinforcing the growth of this branch of the industry.

Electrical manufacturing has contributed a great deal to the industrialization of our nation, and has also made it possible for us to enjoy the benefits of the numerous electrical and electronic devices on the market today. The mass production of appliances and machines has played an important role in providing us with a higher standard of living and more leisure time. The future of electrical manufacturing is virtually unlimited—its potential is beyond the scope of our imagination—yet, through the industry's ingenuity, we are constantly discovering new ways to benefit from this vast field.

With the country's industrialization came a need for workers to fill the many newly established jobs within the electrical industry. One of the areas that developed rapidly was the service, repair, and maintenance of electrical systems and equipment. The workers in this area soon became known as *maintenance electricians*. The nature of their duties has not changed significantly until recently, when they have been confronted with a much greater degree of sophistication.

Within a short time, electric motor repair and electric sign service began to emerge as distinct specialties, and these two branches have become vital segments of the electrical industry today.

CURRENT TRENDS AND FUTURE PROJECTIONS

With increased use of electricity in the United States, it is evident that there will be a continual need for skilled workers in the electrical industry. The construction of electrical generating plants, transmission and distribution systems, industrial plants, commercial and residential buildings, educational institutions, hospitals, and public facilities, and the rebuilding of our country's infrastructure will require a large number of skilled electrical workers.

During the early 1990s the demand for new housing will remain high to meet the needs of our growing population. Business expansion and renovation of existing buildings also will require more building and construction work. The development of energy resources and transportation systems will provide for additional jobs in the construction industry. According to the United States Department of Labor, Bureau of Labor Statistics, employment in the construction sector is expected to increase from about 4.2 million to 5.8 million by the year 2000.

Data released by the Bureau of Labor Statistics indicate that in 1989 there were more than 300,000 construction electricians; employment in this field is expected to increase throughout the 1990s. In addition to this anticipated growth, thousands of job vacancies will arise from the need to replace journeymen who change their employment to another sector within the field, seek employment in other industries, retire or die.

There will be about 6,400 job openings per year to the year 2000 for construction electricians, according to projections of the Department of Labor, Bureau of Labor Statistics.

EARNINGS AND GENERAL WORKING CONDITIONS

Building trades workers receive the highest hourly wages paid to skilled workers, and, within this group, the electrician is quite often the pacemaker. Union wage rates for seven building trades in 33 cities as reported by the Construction Labor Research Council are illustrated in Table 1. The same report reflected a pay range of $17.25–$37.30 per hour for the construction electrician.

TABLE 1

Craft	Hourly Union Wages Summer 1989 Range
Electricians	$17.25–$37.30
Carpenters	$15.00–$34.13
Painters	$12.83–$29.77
Plasterers	$14.29–$31.70
Bricklayers	$15.80–$34.57
Plumbers	$17.86–$45.78
Laborers	$ 9.69–$26.34

Source: Construction Labor Resource Council
Note: The Construction Labor Resource Council is an industry supported wage and cost data research organization based in Washington, D.C. The above data are from a survey of thirty-two cities.

Although the construction electrician receives a substantially higher hourly wage than most other electricians, the work is much more seasonal. Further, because of the nature of construction work, the journeyman usually has limited opportunity to work overtime. The combination of these two factors has had a marked influence on the annual earnings of these workers. Consequently, when one compares the annual earnings of the construction electrician to those of the maintenance electrician, journeyman lineperson, shop journeyman, and electric sign serviceperson, one will note no substantial difference.

Apprentices. Most apprentices are employed on a specified hourly wage basis, as provided in the collective bargaining agreement. An apprentice's wage is usually expressed as a percentage of the journeyman rate. It ranges from about 35 percent during the first six months of employment to approximately 90 percent as the apprentice enters the last six months of the fifth year in training.

Workweek. Generally, the basic workweek consists of 40 hours, the aggregate of a 5-day work week of 8 hours per day. However, there are a number of workers in the industry who enjoy a standard workweek of 35 hours or less.

Premium pay. On most new construction jobs, electricians who are required to work outside the workday stipulated in the collective bargaining agreement receive pay above the regular rate of pay. This premium hourly compensation is commonly known as "OT," the abbreviation for overtime.

Some contracts covering utility, maintenance, and electric motor shop employees specify 1 1/2 times the regular rate of pay for overtime work. Other collective bargaining agreements set the premium rate of pay for overtime at two times the regular hourly rate. Thus, overtime pay rates vary according to the particular branch of the industry, with a range of 1 1/2 to two times the straight time hourly rate.

Vacations. Paid leisure is an accepted part of the American way of life, with longer vacations and more paid holidays being enjoyed by a larger segment of the country's work force. Most collective bargaining agreements covering electrical workers provide for paid vacations. The length of the vacation is usually related to the term of employment, especially in the nonconstruction sectors of the industry. Table 2 illustrates the general trend in vacation allowances at present.

Other Fringe Benefits. Today many workers are covered by health and welfare insurance programs. They also enjoy such benefits as dental and eye care insurance, prescription drug programs, and educational assistance aid. There has been continued growth in the area of private pension plans. Workers realize that there is a need to plan not only for the short term, but also for the future.

TABLE 2 Paid Vacation Benefits
(Electrical Industry Contracts)

Length of Service Required	Average Period of Vacation in Days
6 months	5.0
1 year	7.7
3 years	10.4
5 years	11.9
10 years	15.7
15 years	17.8
20 years	20.1
25 years	21.4

Vacation plans in the electrical construction industry are usually formulated on a cents-per-hour-worked basis. For example, assume that an electrician receives 75 cents per hour vacation pay and is employed for 2,000 hours during the course of the year. The amount of vacation pay the worker would be entitled to is the product of $.75 x 2,000 = $1,500, which in terms of time off would be approximately two weeks' paid vacation.

Compensation for holidays not worked is one of the most common fringe benefits provided for organized workers, according to data reported by the Bureau of Labor Statistics. BLS reports that 62 percent of the workers employed by medium and large firms receive ten or more paid holidays. This benefit is not prevalent in the construction industry, but, as the desire for more leisure time gains impetus, it should begin to receive greater acceptance. In short, the electrical industry provides many lucrative opportunities for those seeking careers in this progressive field.

DESIRABLE ATTRIBUTES AND ASSETS

The electrical trades require a unique combination of talents. Those who enter the field must possess a natural aptitude for using tools and must be able to master the required theoretical discipline. Of even greater importance, the trainee should possess a keen interest and desire for achievement in the field.

Physical Requirements

Apprenticeship applicants should be physically qualified. That is, they should be able to perform the required manual functions in a safe, professional manner. Although much of their work is done indoors, they should be alert, agile, and physically capable of working outdoors under adverse weather conditions.

Being able to drive a lightweight truck or van can be an asset to workers pursuing a career in the electrical field. Also, knowledge of the area's geography in which you will be working is quite beneficial, especially in terms of understanding street maps and determining effectively how to get from one place to another.

Age and Education Requirements

Beginning apprentices in the electrical construction industry are generally between 18 and 24 years old, but allowances of up to four years are made for veterans. A high school education is essential, preferably one with emphasis on mathematics and science. If trainees have a natural inclination for these subjects, they should do quite well in related class instruction because the study of electricity incorporates a wide variety of formulas and mathematical computations and an understanding of chemistry and physics.

Becoming an electrician can be an interesting and challenging assignment, especially for those who enjoy physical as well as mental achievement. In short, a successful career in the electrical field depends largely on the attitude and enthusiasm displayed by the apprentice.

CHAPTER 2

OCCUPATIONAL PLANNING AND SELECTION

Planning a career in the skilled trades involves careful preparation. The successes and failures experienced by many young people are closely correlated to the diligence with which they plan for their vocation. Assuming that you are interested in the electrical field and that you have the potential for meeting the criteria given in chapter 1, you should become acquainted with the various ways in which you can improve your chances of embarking on a successful career in the electrical trades.

EDUCATION AND TRAINING

The educational level of the country's work force will undoubtedly continue to rise during the 1990s, and the importance of educational training cannot be overemphasized. Presently, almost every skilled occupation requires its workers to have a high school education. And even though some industries hire trained high school dropouts, the standards set for promotion prevent upward mobility without a high school diploma.

Education is a highly marketable commodity. It provides the individual with the tools necessary for upward mobility in the workplace. Traditionally, high school graduates have enjoyed substantially greater earnings over their working careers than those who did not complete high school. A certificate from a four-year, accredited high school best prepares a young person for employment. However, if you have not finished high school and are not attending now, there are ways in which you can still get the required training—the equivalent of a high school diploma.

GED Tests

In 1945, a program known as General Education Development (GED) was set up to accommodate returning veterans who had not completed their high school training. The American Council on Education developed a GED test battery which covers the following five disciplines:

1. Correctness and effectiveness of expression
2. Interpretation of literary materials
3. General mathematical ability
4. Interpretation of reading materials in social studies
5. Interpretation of reading materials in natural sciences

Since its inception, the scope of the program has broadened to the extent that many states now allow anyone 17 years or older to participate. Each state department of education establishes the policies and procedures for an adult resident to earn a high school equivalency diploma based on the results of the GED test. The state department of education determines the minimum test scores required to attain a diploma. It also establishes such criteria as age, residency, and any previous high school enrollment requirements for admission to the program. These regulations vary from state to state. If you are interested in this program, contact your state's department of education for specific details about its administration or write to the American Council on Education, One Dupont Circle, Washington, D.C. 20036.

Another alternative to GED programs is correspondence course instruction. A number of reputable institutions provide home study educational programs, and enrollees can do quite well with the courses if they persevere. Students are, of course, primarily responsible for their rate of progress.

Veterans' Benefits

If you are a veteran of the armed forces, you are entitled to education and training under the Veterans' Readjustment Benefits Act of 1966 as amended. This law provides the most important source of aid to veterans. Under it, they may receive benefits while training on the job or going to a high school, trade school, or college. To encourage dropouts to complete their schooling, the 1966 Act, as amended, included a special provision permitting veterans to receive benefits while taking remedial courses leading to the completion of high school without affecting their entitlement to benefits for vocational training or college.

Training may be taken in any approved vocational school, business school, college, professional school, or any establishment providing apprentice or other training on the job. All courses must be approved by the designated state approving agency. The amount of benefit is determined by the following factors: the type of educational/training program; whether it is full-time, three-quarter-time, or half-time; and the number of dependents for whom the veteran is responsible. Under an apprenticeship program or in an on-the-job training situation, a veteran with no dependents would receive monthly benefits of $249 for the first six months, $186 per month for the second six months,

$124 per month for the third six months, and $62 per month for the fourth six months and any succeeding months of training. Cooperative educational/training programs for which GI benefits are provided combine formalized education with training in a business or industrial establishment, with emphasis on the institutional portion.

The Veterans Administration publishes a booklet, "Fact Sheet IS-1," that contains much useful information concerning most federal benefits for veterans, their dependents, and beneficiaries. Also your area regional office of the Department of Veterans Affairs can furnish you with additional details on other special services available to veterans.

EMPLOYMENT SERVICES

In addition to your educational planning, you will want to become familiar with several key sources of useful employment information and assistance. State public employment offices are affiliated with the United States Employment Service of the United States Department of Labor's Employment and Training Administration. These offices provide the potential job seeker service in four pertinent areas: job information, employment counseling, referral to job training, and job placement. Trained personnel give applicants information about the various jobs available, specific job requirements, opportunities for advancement, rate of pay, and other related data. Employment counseling also assists both beginners and experienced workers who wish to change their occupations. The primary purpose of this counseling service is to help job seekers become aware of their actual and potential abilities, their interests, and their personal characteristics.

Often, individuals need additional education and training before they can meet the requirements of certain jobs. The role of the referral-to-training section of the public employment office is to suggest appropriate training for job qualification. The services rendered by job placement personnel are just what the title implies. Specifically, they attempt to place workers in jobs for which they are suited, by maintaining close contact with the job market.

Other Information Sources

Other reliable sources of information and assistance in planning a career are public and school libraries, especially career centers and vocational school libraries which have books, pamphlets, and magazines containing information about different occupations. Many libraries receive publications from the Bureau of Labor Statistics, the business community, and labor unions. Frequently, officials at the local branches of labor unions can provide occupational information relevant to their particular fields. Employers and personnel officers can usually furnish helpful job-related information, as can community service centers which are concerned with employment problems in the area.

ASSESSING YOUR OWN ABILITIES

How do you go about assessing your abilities? One suitable, effective method is to carefully analyze all the qualifications considered essential for admission and achievement in the field you choose. Also, you may ask yourself a number of pertinent questions such as: Is my interest in the occupation genuine, or am I merely looking for employment? What is my attitude toward this type of work? Is it too mechanical, technical, or strenuous? Can I take orders and grasp instructions readily? Will my temperament and personality allow me to work effectively as a member of a crew? How do I feel about the classroom training associated with the job? Do I look at the training as an opportunity or as a requirement? Do I understand what will be expected of me, and do I feel that I will be able to measure up to such standards? This is just a sampling of the questions which you can raise and attempt to answer honestly. When you have analyzed your answers, you should be able to determine whether or not you have a real desire to enter the particular field.

SEEKING THE SERVICES OF A GUIDANCE COUNSELOR

As our nation progresses technologically, the choice of an occupation becomes increasingly difficult. Thus, it often is both desirable and necessary to consult with a counselor about your choice. Faced with an array of job possibilities, you need to learn about those opportunities which will neither frustrate you nor waste valuable time and effort. Vocational guidance counselors can help you recognize your personal characteristics and capacities and so help you determine whether or not you are suited for the particular type of work in which you are interested.

On the basis of the collected data from questionnaires, assessment of personal interviews, results from testing, and other pertinent information, the counselor arrives at a meaningful appraisal of your capacities, aptitudes, personal traits, and occupational inclinations. From this information, the counselor is able to recommend a specific field of employment in which you are likely to achieve success.

CHAPTER 3

DEVELOPING YOUR ABILITIES THROUGH EDUCATION AND TRAINING

Getting the most from preparatory education requires foresight and planning. You must seek the type of training that will allow you to develop a solid base on which to cultivate your talents.

Almost any shop or laboratory course offered in junior or senior high school will be beneficial. Such courses will acquaint you with the correct use of basic hand tools and simple shop equipment. They also help you learn work practices.

Perhaps the most prominent feature of this training is that greater emphasis is placed on developing attitudes and stimulating interest than on skill improvement. You are able to explore your desires and, at the same time, evaluate your aptitude for the type of work.

The basic courses should be followed by in-depth courses that allow you to develop fundamental skills and broaden your foundation for future growth and specialization.

PREREQUISITE COURSES

An electrician is often called on to perform complex tasks that require knowledge of science and technology. Trainees should prepare themselves by taking courses which further their education in science and mathematics. Courses in these disciplines help individuals grasp subsequent instruction on a more sophisticated level.

Apprentices endeavoring to learn the electrical trade are confronted with many formulas, equations, and computations that necessitate a thorough understanding of mathematics. Courses in mathematics that should prove beneficial to future electrical workers include algebra, trigonometry, and geometry. A study of physics and chemistry also lends itself favorably to the study of electrical theory, since it relates directly to these subjects.

In addition, individuals seeking a career in the electrical field will do well to take courses in electronics, air conditioning and refrigeration, welding, mechanical and electrical drawing, blueprint reading, estimating materials and supplies, technical report writing, and industrial safety. Many educational institutions, having recognized the need for instruction in these areas, have established specific objectives for each of these courses.

Prerequisite courses should be flexible. They should provide training that is transferable to and usable in different occupations. It is also important that the student pursue these courses in logical sequence.

INDUSTRIAL ARTS

The underlying purpose of industrial arts in junior and senior high schools is to acquaint students with the industrial environment in which they must live and work. A brochure entitled

"Industrial Arts in Education," published by the American Vocational Guidance Association, describes industrial arts as the study of industrial tools, materials, processes, products, and occupations pursued for general education purposes in shops, laboratories, and drafting rooms.

There are, of course, specific training objectives involved in an industrial arts program. Some of the more significant of these goals are: (1) to promote clear analytical reasoning when performing mechanical and constructive tasks; (2) to develop efficient work habits; (3) to cultivate favorable attitudes toward group activities in performing constructive enterprises; (4) to develop an understanding of and an appreciation for industrial materials, processes, and products; (5) to cultivate a wide variety of skills in the use of tools and machines; and (6) to promote safe work habits and proper attitudes toward fellow workers.

Industrial arts education is an element of general education. The specific design of an industrial arts program is not directed toward tangible vocational goals. However, these programs often serve to inspire students who have a desire to create and construct projects through the coordination of their minds and hands. Meaningful industrial arts curriculums can provide an excellent background for young people who intend to pursue a vocational education. Industrial arts also allow an individual to see an industry, such as construction or service and repair, as a whole, thus providing the person greater insight into the composition of that industry.

VOCATIONAL AND INDUSTRIAL EDUCATION

What is vocational education? In brief, it is a program of specialized education designed to provide instruction leading to occupational competency. The term "vocational education" applies to all forms of education and training that are designed to prepare people for successful employment. Vocational-industrial education includes trade training for industrial occupations.

Vocational-industrial education, like industrial arts, involves educating the whole person. Thus, it complements, rather than competes, with general education. However, unlike industrial arts education, it places major emphasis upon developing salable skills and knowledge. Vocational-industrial education curriculums prepare students for specific occupations, provide industry with a source of trained workers, and provide training and retraining for adults who are unemployed or underemployed.

Recently, there has been a change in the level at which vocational education is being taught; the trend is toward teaching vocations at the post-high-school level. This is contrary to past procedure, when many vocational institutions operated below high school level.

Because of the upward movement of the general education level in the United States and the advancement of technology and automation, the upgrading of vocational-industrial education was an obligation that had to be fulfilled. Today, many fine trade and technical institutes throughout the nation offer excellent facilities and programs which provide students with the opportunity to pursue occupational training relevant to their needs and desires.

The redirection of the public vocational education program set in motion by the Vocational Education Act of 1963, and moved forward by subsequent amendments, should greatly strengthen the occupational preparation of millions of young people not bound for college. Specifically, these amendments emphasize vocational training closely related to current job markets.

Federally aided vocational education programs ease the school-to-work transition by providing occupational training oriented to job market needs. Cooperative education is one type of vocational education aimed at easing the school-to-work transition problem. In this type of program, a formal relationship is established between the high schools of an area and the vocational education institute serving that community. Students who participate in this program are allowed to divide their time between the two school systems. Their high school curriculum broadens their general education, while their attendance at the vocational school affords them the opportunity to receive occupational training. Another form of cooperative education is that which divides a person's time between the secondary or postsecondary educational institution and employment.

The figure below shows the structure of vocational trade and industrial education within the framework of many of our present school systems.

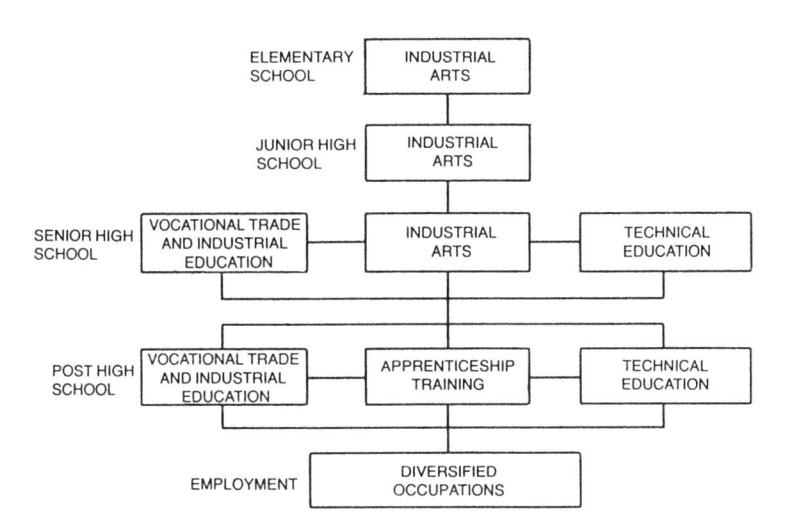

28   *Opportunities in Electrical Trades*

SAMPLE APPRENTICESHIP SCHEDULES

The following schedules were extracted from the U.S. Department of Labor Employment and Training Administration's Trade and Industry Publication No. 4, 1970 edition. It is important to note that these schedules are still valid. They describe, in part, the job contents for selected occupations within the electrical industry. These outlines are included in this chapter to acquaint the student with a sampling of the work experiences and training associated with various electrical occupations. These outlines are intended solely as guides.

Construction Electrician

I. Residential . . . . 2,000 hours
    Services—single phase
    Metering—single phase
    Remodeling
    Installation of conduits
    Installation of B.X. cables
    Installation of outlets
    Installation of special equipment
        Hot water heaters
        Electric ranges
        Exhaust fans
        Garbage disposal
        Electric heaters
        Heating systems
        Annunciator systems
        Door bells
    Installation of light fixtures, receptacles, and switches
    Security systems

II. Commercial-Industrial . . . . 5,000 hours
    Services
        240 volt
        480 volt
        Over 600 volt
    Metering
        Polyphase
        Current transformers
    Installation of conduits and outlets
        Concrete slab and masonry
        Steel construction: exposed, concealed
        Buss-duct systems
        Under-floor duct systems
        Metal raceways and troughing
        Explosion proof
        Vapor proof
        Flexible conduit and cables
    Circuiting
        Three-phase circuits—light and power
        Feeder circuits—light and power
        Branch circuits—light and power
        Control circuits—light and power
    Various types of motors and their controls
    Transformers—application and connection

III. Specialized work . . . . 1,000 hours
    Welding
        Acetylene
        Electric arc
        Acetylene burning
    Management-employee relations
    Customer-employee relations
    Electronics systems

Communications systems
Fire alarm systems
TOTAL . . . . 8,000 hours

Industrial Maintenance Electrician

I. Electrical construction . . . . 2,076 hours
    Safety instructions
      Bend and install conduit
      Bend and install other wiring
      Run wire and make hook-ups
      Install lighting and power circuits
      Install power and control equipment
      Install and line up motors and generators
      Lay out job from blueprint and select material

II. General maintenance . . . . 1,766 hours
    Safety instructions
      Check lights
      Electrical equipment
      Machinery
      Lubrication
    Diagnose trouble in lighting and power circuits
    Locate cabinets and distribution boxes
    Periodically check and repair electrical equipment
    Adjust and repair welders

III. Cranes and elevators . . . . 520 hours
    Safety instructions
      Check, repair, and adjust limit switches
      Check, repair, and adjust saftey devices
      Locating and repairing faulty electrical equipment

IV. Electrical repair . . . . 1,650 hours

Safety instructions
  Motor repair (A.C. and D.C.)
  Test block
  Controller repair (A.C. and D.C. starters)
  Heating appliances
  Building and repairing transformers and welders
  Drill repairs

V. Powerhouses: Substation construction . . 850 hours
  Safety instruction
    Heavy cable installation
    Lay out and install heavy conduit and duct
  Install master distribution cabinets
  Lugging, rubber covered cable splicing, lead cable splicing, installation of pot heads
  Connect cable in master and distribution cabinets
  Wiring busways and bus
  Wiring switchboards and switchgears
  Wiring transformers, motors, and generators
  Connect instrument transformers, meters, and relays
  Check and test circuits

VI. Maintenance of powerhouse and substations . . . . 466 hours
  Safety instructions—servicing of main and auxiliary circuits
  Servicing of equipment
  Adjusting of the timing of relays
  Testing and reconditioning of the transformer and oil switches, coolants for transformers and switches
  Testing of all circuits

VII. Related instruction . . . . 672 hours

TOTAL . . . . 8,000 hours

32 Opportunities in Electrical Trades

Electric Motor Shop Journeyman

I. Shop work . . . . 8,000 hours
    Shop errands and orientation
    Cleaning electrical machinery and auxiliary equipment of various kinds
    Stripping old windings, cleaning bar coils and tinning leads
    Appliance and fan repairs, locating trouble in fractional H.P. motors (A.C.) and repairing
    Winding field coils, D.C. armatures, transformers, coils, and solenoid coils. Test field coils and armatures for shorts, opens, and grounds.
    "Troubleshooting" on various types of equipment in the shop.
    Winding and forming armature and stator coils, also redesign of same
    Commutators and general rebuilding of same
    Batteries and battery charging equipment
    Starter and control repairs; wind and repair armatures, rotors, and stators; repair brushes, brush holders, and motor leads
    General machine shop practices
    Assist with trouble shooting in field
    Switchboard work
    Welding—spot and flame
    Taping and insulating coils, also bar windings
    Testing various types of electrical apparatus

TOTAL . . . . 8,000 hours

Marine Electrician

I. Wireways . . . . 1,000 hours

Education and Training 33

        Layout of raceways—raceway prints
        Making and installing hanger—tubes
        Cutting threads, bending and installing conduit
        Installing cables
        Stripping cables

II. Power wiring . . . . 2,000 hours
        Print reading
        Installing motors and controllers
        Connecting motors, controller switches, circuit breakers

III. Marine lighting . . . . 500 hours
        Layout print reading
        Installing fillings and fixtures
        Making joints, soldering, taping and splicing

IV. Interior and other communications . . . 1,000 hours
        Bells, batteries, alarm connections, and gyros
        Telephones
        Engines, order telegraphs; follow-up system
        Radio installations
        Indicator—prints

V. Installation of generators and switchboards . . . . 500 hours

VI. Yard maintenance . . . . 2,000 hours
        Lights, transformers and connections
        A-C power, motors, cranes

VII. Operating electrical equipment . . . . 1,000 hours
        Tests—circuit continuity
        Megger, voltage, amperes, measurements

TOTAL . . . . 8,000 hours

Electrical Lineman (Light and Power)

I. Prerequisite experience—loading and unloading tools and supplies on trucks; setting poles and anchors; sending tools and supplies to lineman on pole; other assigned work on ground

Names and uses of tools and materials
Care of tools and rubber goods
How to tie rope knots and slings
How to make various partial assemblies of materials on the ground before installation on the pole

II. Line construction: setting, guying, and anchoring poles, and pole-top work

Use and care of climbing tools
Pole climbing skill
Safety precautions, including use of rubber gloves and blankets
First aid and safety methods
Team work, sequence of operations
Public relations and safety
How to arrange for short service interruptions
How to keep job and work order accounts
Pole line construction methods: span lengths and sagging-in, hardware
Tree clearance
Temporary construction schemes to avoid service interruptions
Principles of electricity
Reading electric drawings and sketches, and ability to make a legible sketch or diagram

III. Building transformer structures, installing transformers and protective equipment, and connecting to lines

Construction methods for transformer structures
How to connect and phase out transformers and lines
Knowledge of connections commonly used
How to work on energized lines up to 4,500 volts

IV. Street lighting installation
Series and multiple circuits
Methods of making installations

TERM . . . . 4 years

Powerhouse Electrician (Light and Power)

I. Orientation
Learn physical location of electrical equipment
Familiarity with power plant terminology
Knowledge of company organization
Familiarity with symbols and switching diagrams

II. Maintenance of electrical equipment, including oil circuit breakers, lighting equipment and starters
Knowledge of routine maintenance required, such as oil changes, cleaning, and inspection
Learn proper oils and greases to be used for specific pieces of equipment
Know how to remove electrical equipment from service and obtain clearances before switching
Proper care of storage batteries
Knowledge of simple lighting circuits and methods of repairing lighting fixtures

III. Check and maintain safety equipment
Familiarity with various types of fire extinguishers and methods of refilling
Knowledge of first aid and safety methods

Proper use of rubber gloves and blankets
Knowledge of all types of fuses and fuse cartridges in use
Understanding of methods of fighting electrical fires

IV. Perform shop and plant duties, as required

Care and use of hand tools
Knowledge of soldering, brazing, and elementary welding (gas and electric types)
Care and use of power tools (lathe, drill, press)
Knowledge and use of insulation materials, and knowledge of checking insulation by use of megger
Use and interpretation of a "growler"

V. Maintain records, as required
Knowledge of purpose of records
Basic accounting for proper distribution of materials

VI. Install conduit and wiring for any electrical equipment

Method of installing conduit
Knowledge and proper use of conduits
Familiarity with the use of proper cable and wire as pertains to size and insulation
Method of handling wire in conduit
Knowledge of conduit sizes and permissible circuits in any size

VII. Install and connect transformers

Methods of installing and connecting transformers
Knowledge of connections and circuits of current, potential, and power transformers
Theory and safety precautions to be observed
Phasing between banks
Checking voltages of various connections

Education and Training 37

VIII. Installation and maintenance of regulators
>   Theory of induction and voltage regulators
>   Method of control
>   Routine care, methods of installation, and maintenance

IX. Installation and maintenance of rotating machinery
>   Theory, care, and installation methods of A.C., D.C. generators and motors (includes single-phase, polyphase, synchronous and induction constant and variable speed for A.C. and series, shunt, and compound wound D.C. machines)
>   Routine care
>   Methods of checking motor loads
>   Adjustments and repairs of starters
>   Methods of checking motor clearances and motor bearings
>   How to reconnect for different voltages and speeds
>   How to care for commutators
>   How to locate and cut out defective motor coils (A.C. and D.C.)

X. Install meters, relays, and controls
>   Basic theory of watt-hour, meter, voltmeter, ammeter, and wattmeter
>   Analysis of control diagrams
>   Knowledge of relays and remote controls
>   Adjustments of elevator controls
>   Knowledge of proper uses of voltmeters and ammeters
>   Understanding of interlocking controls
>   Knowledge of operation of industrial instruments such as temperature recorders and remote flow meters

38   Opportunities in Electrical Trades

>
> Knowledge of location of grounds and short circuits and ability to apply emergency correction
> Understanding of when to use "stay put" or other types of push buttons, as well as no-voltage releases, for continuity of service

XI. Perform general plant construction

> Working knowledge of plant power distribution
> Knowledge of wiring diagrams
> Blueprint reading
> Elementary knowledge of electronics and its industrial use
> Methods of laying out general plant construction work
> How to plan and choose switch capacities, circuit capacities, controls, and plant installations

TOTAL .... 8,000 hours

Sign Electrician

I. Shop wiring

> Primary insulation
> Secondary insulation
> Flashers
> Transformers
> Time clocks
> Switches
> Fuse blocks

II. Shop assembly

> Installing tube supports
> Bushings
> Housings

Mounting tubing
III. Shop erecting
  Prefabrication
    Structural steel
      Roof signs
      Marquee signs
    Operations required
      Burning
      Punching
      Welding
      Bolting
      Coping
      Loading
IV. Job erecting
  Layout
  Preparation of site for installation
  Erection of steel structure when necessary
  Installation of rigging
  Landing
  Securing signs
  Connection to power supply
V. Service or maintenance
  Troubleshooting
  Removing defective parts
  Replacing defective parts
  Repairing, patching, and refinishing

TERM . . . . 4 years

APPRENTICESHIP TRAINING

The individual worker's degree of skill determines the quality of work he or she can do. To develop skills fully, the worker must receive training, and an excellent way to obtain it is by becoming an apprentice.

An apprenticeship is the recognized method of learning a skilled trade. In the electrical industry, the vast majority of journeymen have acquired their knowledge and skills through apprenticeship programs. The International Brotherhood of Electrical Workers and the National Electrical Contractors Association recognized the need for pertinent apprenticeship training standards for the electrical construction industry in 1941 and formed a National Joint Apprenticeship and Training Committee. This committee was assigned the responsibility of formulating training standards, in cooperation with what is now the Bureau of Apprenticeship and Training, United States Department of Labor. Because of the dynamic nature of the electrical industry, it is necessary to review and revise these standards from time to time.

The effectiveness of a training program depends, in large measure, on the training techniques involved. Apprenticeship training combines on-the-job training with related classroom instruction, reinforcing the learning process. In current practice, it provides a job for the trainee while providing an opportunity to acquire knowledge and develop skills. Most electrical apprenticeship training programs are based on a five-year curriculum.

The role of the Bureau of Apprenticeship and Training, a part of the U.S. Department of Labor's Employment and Training Administration, is to assist labor and management in the development, expansion, and improvement of apprenticeship and training programs. The bureau's principal functions are to encourage the establishment of sound apprenticeship and training programs and to provide technical assistance in setting up such programs. The bureau works closely with state apprenticeship agencies, trade and industrial education institutions, and labor and management. Additional information on apprentice programs may be obtained from the local office of your state's employment service.

The importance of apprenticeship training can not be overestimated. Former Secretary of State Ray Marshall, who served as Chairman of the Federal Committee on Apprenticeship, said in an address to the New York State School of Industrial Labor Relations, "...that an expanding and improving apprenticeship system is essential to the welfare of workers and the economic health of the nation...."

CHAPTER 4

OCCUPATIONS IN THE ELECTRICAL FIELD

Electrical workers, in general, fall within two major industry divisions as defined by the Bureau of Labor Statistics. These two divisions are the goods producing and the service producing sectors. The construction and manufacturing industries are part of the goods producing sector. Transportation, utilities, and other services come under the service producing sector. In 1988, these industries provided for a total of over 55 million jobs; they are projected to encompass nearly 65 million jobs in the year of 2000. These industries provide the bulk of job opportunities for electrical workers and other skilled occupations. Skilled workers have higher earnings, more job security, better chances for promotion, and more opportunities in general than most workers without a skilled occupation or professional background.

In 1986, according to BLS, there were 13.9 million precision, production, craft, and repair workers. Employment for this important group of workers is expected to reach 15.6 million by the year of 2000—an increase slightly in excess of 12 percent. In this book, we will limit our discussion to selected occupations within the electrical field. BLS reported that in 1988 there were 542 thousand electricians employed and that this group of workers could reach 638 thousand in the year 2000.

The electrical industry encompasses a multitude of occupations and job classifications, and space does not permit the inclusion of all the occupational titles that are relevant to the electrical industry. For a comprehensive listing of job titles, the reader may want to refer to the *Dictionary of Occupational Titles*, published by the U.S. Department of Labor, Employment and Training Administration, Bureau of Employment Security. It can be found in most public libraries and in career centers in high schools. The occupations selected here were chosen because they are key jobs and mirror a wide cross section of job content and skills.

Various electrical jobs have common elements. A qualified journeyman usually has a degree of job mobility within his or her field. The worker may decide to work in a job classification other than his or her normal one; for example, a construction electrician may seek employment as a maintenance electrician, or a maintenance electrician might look for employment as a marine electrician. Within reasonable limits, there are other interchangeable combinations.

CONSTRUCTION ELECTRICIAN

The duties of an electrician are numerous and varied, presenting an exciting challenge to individuals seeking a career in this field. *Inside construction electricians* lay out, assemble, install, and test electrical circuits, fixtures, appliances, equipment, and machinery. They must be able to install electrical lighting, heating, cooling, and control systems in various types of structures: residences, commercial and industrial establishments, schools, hospitals, libraries, and other buildings. They also are frequently called on to make sophisticated installations involving electrical motors, controllers, transformers, switchgear, and other electrical apparatus.

To cope with their tasks, construction electricians must be able to apply the sciences learned during apprenticeship training. They also must be able to use diverse electrical formulas and computations associated with their work, such as determining the size of electrical service conductors, feeders, and branch circuits.

In complying with the requirements of the National Electrical Code and with other safety standards, electricians often must make calculations which enable them to determine the allowable number of conductors a certain size conduit or raceway will accommodate safely. They also must install interior wiring circuits which include the circuits' conduit or raceway system, the wiring or conductors, fixtures and appliances, and protective devices which ensure safe operation of the system.

In addition to wiring buildings and industrial plants, construction electricians install street lighting systems, motorized equipment for bridges, machinery and wiring for power plants and substations, and sophisticated communications, alarm, and security systems. They also carry on a host of other projects that fall within the domain of the building trades. They must be familiar with the National Electrical Code, which sets forth acceptable standards governing electrical installations. Electricians also must be aware of the existing state, county, and municipal regulations pertinent to their work.

Electricians furnish their own hand tools: screwdrivers, pliers, levels, plumb bobs, hammers, pocket knives, hacksaw frames, compass saws, braces and bits, pipe wrenches, and adjustable wrenches. An initial expenditure of $275 will usually provide a construction electrician with an adequate supply of hand tools. The electrical contractor supplies large tools and equipment, including hydraulic benders, power tools, ladders, and such expendable items as hacksaw blades, taps, and twist drills.

Unlike the other occupations to be discussed, employment for the construction electrician is somewhat seasonal. There usually are more employment opportunities in the summer than during the winter season, especially in parts of the country where the winters tend to be severe.

Employment practices for the construction electrician also differ in another way. Whereas the maintenance electrician, electric motor shop journeyman, marine electrician, utility and power plant electrician, and electric sign serviceperson change employment infrequently, construction electricians often work for a substantial number of contractors during their careers.

Most construction electricians work for electrical contractors. Some work for government agencies or business establishments that do their own electrical work, and a few are self-employed.

The work of the construction electrician, like that of other building trades workers, is active and requires a moderate degree of physical strength. It also calls for an alert and attentive mind. Since a large measure of the work is performed indoors, the inside electrician is not subjected to the elements as often as other building trades workers.

Construction electricians are crafts workers, and like all genuine skilled workers, they derive satisfaction and pleasure from utilizing both their minds and hands in a constructive manner.

MAINTENANCE ELECTRICIAN

Basically, the major differences between the construction electrician's duties and those performed by a maintenance electrician lie in the application of their individual skills. The maintenance electrician is responsible for maintaining existing electrical systems, equipment, and machinery. The electrician must periodically service, and, when necessary, make repairs that ensure the continuous operation of the facility in which he or she works. Many firms employing maintenance electricians require their personnel to establish a preventive maintenance program. This program sets forth a specific time schedule to be followed in servicing the plant's machinery and equipment. Servicing machinery and such equipment as motors, generators, controllers, switchgear, and other electrical apparatus normally entails the following processes: cleaning, minor adjustments, replacement of worn parts, and any other necessary repairs.

Many companies operate 24 hours a day. Consequently, they must utilize a work force consisting of three shifts, each electric crew working eight hours per day. Normally, most of the maintenance personnel are assigned to the day shift, which performs the major part of the routine maintenance. However, in order to ensure continuous operation, the second and third shifts also include maintenance crews. Maintenance electricians employed on the second and third shifts are responsible for preventing costly production losses and inconvenience. In emergencies, they may advise management whether or not an immediate shutdown of equipment is necessary, the amount of time needed to repair the equipment, and whether continued operation would be hazardous.

The maintenance electrician's daily work includes tasks other than routine maintenance assignments and performance of emergency repairs. The maintenance electrician must be able to plan, assemble, install, and test electrical circuits and systems which frequently include complex sensory, monitoring, and other sophisticated control devices. The work may also include the installation of lighting circuits, distribution and load centers, fuse panels, magnetic starters, safety switches, electric motor controllers, rectifiers, electrical machinery and equipment, and many other types of electrical apparatus. Thus, the worker must also be able to interpret and work from blueprints and wiring diagrams.

Like the construction electrician, the maintenance electrician must be familiar with the National Electrical Code and any guidelines set forth by other pertinent regulatory agencies which control installation practices for the specific geographical area. Since a maintenance electrician almost always preforms duties for one organization, he or she is usually not required to have the same type of certification as the construction electrician. Licensing and certification will be discussed in a subsequent chapter.

Whenever a piece of electrical equipment or machinery fails to operate properly, the maintenance electrician must locate the trouble. The technique employed in locating circuit, equipment, or machinery faults is known as *troubleshooting*. To help diagnose troubles effectively and expediently, the maintenance electrician utilizes a variety of test equipment. The following are examples of some of the instruments and test equipment used fairly often: voltmeters, ammeters, voltage testers, test lamps, oscilloscopes, wattmeters, power factor meters, and instrument transformers, logic probes, and various digital devices.

In addition to the tools owned and used by the construction electrician, the maintenance electrician's tool chest customarily includes box and open-end wrenches, a ball-peen hammer, punches, callipers, and a socket set. The complement of tools may range from $300 to $350 in value.

Employment for most maintenance electricians is rather stable. Often, they are employed by only one firm during their entire working career. A large measure of their work is performed inside, and they are seldom subjected to severe weather conditions. But because of the nature of their work, they must frequently work in close quarters and may be exposed to unpleasant indoor temperatures.

Maintenance electricians enjoy and take pride in their work, and they characteristically display a high level of ingenuity. The hourly pay for maintenance electricians is generally less than that of construction electricians. However, because of their steady employment, maintenance electricians frequently have greater annual earnings.

THE ELECTRIC MOTOR SHOP JOURNEYMAN

Motor shop journeymen are highly skilled artisans. The composition of their work demands that they possess a wide variety of mechanical skills in addition to their electrical competence. Many of the tasks they perform require the use of various types of machines. Shop journeymen must know how to skillfully operate such shop machinery as metal lathes, static and dynamic balancing equipment, drill presses, power saws, hydraulic presses, and an assortment of coil-winding equipment. For this reason, it is essential that a portion of their training include general machine shop practices.

Motor shop mechanics are primarily responsible for the repair and service of electrical machinery and equipment. Basically, their work can be divided into two general categories—work performed within the shop and duties discharged outside the shop. The first applies to all jobs worked on inside the employer's establishment, and the latter covers those tasks executed on the customer's property. Typical shop work includes rebuilding electrical motors, generators, starters, and controllers; rewinding transformers, relays, magnetic brake coils, and other miscellaneous coil windings; and the fabricating of switchboards and panelboards. Often, rebuilding electrical motors and generators entails, in addition to rewinding the major elements, either the replacing or repairing of mechanical parts such as the shaft of the revolving member (armature) or the bearings which support the armature. In order to accomplish these tasks, the shop journeyman must draw upon experience with the use of such shop equipment as the lathe, drill press, hydraulic press, and other associated machines. The lathe can be used to resurface the shaft of the armature and to make replacement bearings, while the hydraulic press is employed to insert newly machined bearings.

Welding, both gas and electric, is a valuable "tool of the trade" to the motor shop journeyman. Welding skills come in handy when there are broken metal parts in need of repair. Also, the process of welding is very useful in the construction of switchboards, panelboards, and other apparatus.

In addition to being mechanically inclined, shop journeymen must be adept at troubleshooting. That is, they must be able to analyze control circuits quickly, diagnose problems in electrical machinery and equipment readily, and make necessary repairs efficiently. A substantial part of the job hinges on the ability to read and understand complex control wiring diagrams and controller and starter wiring schematics.

To assist in the work, the motor repairman utilizes a number of instruments. Commonly used are the voltmeter, ammeter, tachometer, megohmmeter, frequency meter, and wattmeter. These instruments are invaluable to the shop mechanic. They facilitate the process of locating faults in electrical circuits, machines, and equipment. Proper use and knowledge of test equipment is of the utmost importance to those who want to become expert troubleshooters. Hence, there is much similarity between the maintenance electrician's duties in locating and repairing faults in electrical machinery and equipment and those of the shop journeyman which come under the heading of troubleshooting. However, the expertise needed to perform intricate, internal repairs on electrical machinery usually calls for an electric motor specialist—the shop journeyman. Since these persons are also responsible for making installations involving new apparatus, machinery, and equipment they are called on to repair, they must be aware of the electrical codes and regulations pertaining to their area of work.

A large part of the shop mechanic's duties involve rewinding electrical motors, generators, and transformers. Sometimes the apparatus is rewound in place, especially when the equipment is too large or extremely difficult to move. But more often than not, the equipment is removed from its installed position and transported to the repair shop, where it can be repaired much more efficiently. Motor rewinding requires considerable deftness and the ability to concentrate on complex winding arrangements and connections. Proficient armature and motor winders are a select group; their talents are always in demand.

In order to perform the variety of tasks assigned to them, the shop journeymen must own a fully equipped toolbox. Their set of tools costs approximately $375, and the tools used most frequently are the ball-peen hammer, rule, socket wrenches, calipers and micrometers, an assortment of punches and chisels, pocket knife, combination wrenches, pipe pliers, pipe wrenches, hacksaw, hollow-head wrenches, diagonal and side-cutting pliers, screwdrivers, rawhide mallet, and many special winding tools.

There are times when shop journeymen must work on outside jobs. For the most part, however, their duties are performed under shelter. The physical aspects of the work require strength and endurance, as they must often lift heavy machine parts and do a considerable amount of bending.

The shop journeyman's work is interesting and challenging, and demands a high degree of versatility and skill.

MARINE ELECTRICIAN

Our country's ships and other waterborne facilities are kept lighted and electrified by the efforts of the marine electrician.

Almost every operation aboard our modern ships requires electrical power for its accomplishment, and on many ships propulsion is provided by electrical motors which, in turn, receive their energy from electrical generators interconnected through intricate wiring systems. Communications and navigation equipment, pumps and auxiliary machinery, steering gear motors, and a host of other electrical devices all depend on electricity for their operation—and each contributes to the overall performance of a vessel. All this apparatus is connected by miles and miles of electrical cables which are also installed and repaired by marine electricians.

The reliability of a ship's electrical system depends, in large measure, upon the quality and type of maintenance program employed. And the implementation of these programs constitutes a substantial amount of marine electricians' work.

Marine electricians must be versatile. In addition to installing electrical systems, they must be adept at modifying existing systems. Quite often, because of new engineering designs or changes in manufacturing processes, an exact replacement for a specific piece of equipment may not be available. Thus, the marine electrician must be capable of evaluating the task at hand, recommending specific compensating adjustments, and carrying out the requisite duties. Like other workers who service machinery and equipment, the marine electrician must possess a natural ability for mechanics. Locating grounds and other faults in electrical systems is another phase of the work, especially if the employer specializes in ship repair and service. The characteristics of a successful troubleshooter in this field are similar to those required of the maintenance and motor shop electrician.

Because of the nature of their work, marine electricians must be familiar with the various installation techniques unique to their field. They must be aware of the importance of maintaining watertight seals throughout vessels; therefore, it is important that they possess a thorough understanding of the proper and approved methods for installing cables and fittings throughout a ship. Marine electricians must also be familiar with the United States Coast Guard regulations pertinent to their work.

Many marine electricians find employment in the oil industry's offshore drilling operations. These electricians are hired to service and repair the electrical wiring and apparatus aboard offshore rigs. Occasionally, they are required to make new installations or modifications to the existing electrical system.

The marine electrician usually owns a set of tools comparable to those of the maintenance electrician. In addition, however, he or she must own several other special tools used exclusively by marine electricians. Two such special items are cable skinners, used to remove protective armor from cables, and packing tools, used to pack watertight terminal glands.

Marine electrical work requires frequent bending, kneeling, climbing, and standing. The work, at times, requires heavy lifting. Good health and physical strength are essential for success in this field. Marine electricians work both indoors and outdoors, so they must be able to tolerate varying weather conditions. Sometimes they are required to work on deck jobs in freezing weather, and at other times the job must be done in hot engine rooms.

Marine electricians usually have stable employment. Their work is interesting, exciting, and challenging. Ample opportunities for employment in this field should arise as our nation faces the task of replenishing our merchant marine fleet.

UTILITY AND POWER PLANT OCCUPATIONS

Utility and power plant workers are responsible for the generation, transmission, and distribution of electrical power to our communities. These workers operate the numerous power stations spread across our nation; they provide the personnel needed to install, service, and maintain the vast array of complex wiring systems that deliver electrical energy for our instant use.

Many types of workers are needed to produce, transmit, and distribute electrical power and to maintain and repair the machinery and equipment used by utility companies—so there is a variety of job classifications within these groups. However, this discussion will be limited to those occupations that require knowledge and skills parallel to those of other electrical workers. They include: the power plant electrician (maintenance), lineperson, troubleshooter, and cable splicer.

The power plant is the heart of any electric power system. It houses the machinery and equipment used in the conversion of energy from one form to another. Electricity is generated primarily in steam-powered plants. Steam-generating plants use coal, gas, oil, or nuclear energy for fuel. Approximately 80 percent of the energy resources used in this country are fossil fuels (coal, oil, and natural gas). The most critical fuels in terms of estimated reserves are oil and natural gas. Of the total energy-producing resources consumed, approximately 25 percent are used to generate electricity.

Hydroelectric generation is another major source of electrical power, and nuclear fuel is certainly an important source of energy in the production of electricity. Today, in power plant construction, the trend is toward building larger and more fully-automated power stations. Nevertheless, a considerable number of workers are needed to service and repair the machinery and associated apparatus used in the production of electricity.

Power Plant Maintenance Electrician

The power plant maintenance electrician is one of the key persons included in this group. The duties are similar to those of the maintenance electrician. The one difference is that the power plant employee works on equipment that is either directly or indirectly responsible for the generation of electrical energy.

Power plant electricians use essentially the same tools used by other maintenance electricians. Their work is generally performed indoors, on a routine basis, and a substantial portion of the job centers around preventive measures which help ensure against major and costly breakdowns. The power plant electrician must be able to make quick, accurate decisions in carrying out assignments, especially when faced with an unforeseen combination of circumstances calling for immediate action.

Power plant electricians, like other power company personnel, usually work on a shift basis, since supplying electricity to residences, commercial establishments, and industrial plants is an around-the-clock operation. The personal qualities fundamental to success in this occupation are the same as those given for the maintenance electrician.

Once electricity has been generated, it is ready for use instantly. To accommodate and facilitate its usage, vast networks of electrical conductors are employed to convey power to almost every locality in the United States. Transmission and distribution departments of utility companies are responsible for the installation and maintenance of these systems. To discharge their responsibilities effectively, these two departments generally employ a substantial number of workers. Their duties range from driving trucks and operating pole-setting equipment to supervising complex installations. The purpose here is to explore the duties of three principal classifications of workers: the lineperson, the cable splicer, and the troubleshooter.

Lineperson

These workers make up the largest single occupation in the utility industry. Line construction—the setting, guying, and anchoring of poles and structures, and the stringing of high tension lines—constitutes a large portion of the lineperson's duties. He or she must be familiar with pole line construction methods and the terminology associated with this type of work. The job also includes building transformer structures, installing transformers and protective devices, and connecting this equipment to the power lines. Consequently, the worker must possess a thorough knowledge of transformer theory and connections.

A lineperson often works from aerial baskets, which are insulated "buckets" attached to a boom that is mounted on the bed of a truck. The aerial basket or bucket facilitates overhead work, since it can usually be readily positioned near whatever is to be worked on. In some power companies, a lineperson may specialize in particular types of work. Some may be assigned to new construction only, while others may be assigned to crews strictly responsible for repair work.

Frequently, a lineperson is required to perform tasks at critical heights and under hazardous conditions. Also, there are times when he or she must work on energized power lines and equipment under foul weather conditions. When wires, cables, or poles break, it means an emergency call for a line crew, who must often brave the aftermath of a hurricane, storm, or tornado to reestablish the electrical service we have become dependent upon.

Troubleshooter

A troubleshooter is an experienced lineperson assigned to the special crews that handle emergency calls for service. Since a troubleshooter is basically an experienced lineperson, he or she should possess all the skills and knowledge demanded of a proficient lineperson. In addition, this worker should be very familiar with the power company's transmission and distribution network. Troubleshooters often work alone when they are attending to service calls. They are usually assigned a service truck which is adequately equipped to handle the routine calls they receive. They move from one assignment to another, as ordered by a central office that receives reports of line trouble. Normally, the troubleshooter is radio-dispatched to problem areas as they occur. There are times when the worker merely disconnects or isolates the defective component or circuit from the rest of the system, allowing safe operation of the system—or at least a part of it—until a repair crew can be sent to correct the breakdown. Of course, the specific action taken by the troubleshooter depends largely on the nature and extent of the breakdown.

Other duties of the troubleshooter include the following: (1) maintaining overhead and underground primary lines and equipment, (2) using hot-line tools and rubber protective equipment when working on energized lines and apparatus, (3) testing, re-fusing, or replacing transformers, fuse cutouts, and lightning arresters to restore or maintain service, (4) clearing shorts or grounds from lines, (5) restoring downed wires and trimming trees that create a hazard to primary or secondary wiring, (6) making temporary repairs to poles, guys, and fixtures, (7) informing supervisors of repairs requiring the services of a line crew, (8) clearing trouble at unattended or automatic substations by switching, re-fusing, and setting regulators, (9) patrolling and observing the condition of lines, poles, insulators, transformers, cutouts, and other distribution equipment, (10) maintaining street lighting systems, and (11) replacing broken or burned-out street lamps. They also perform routine clerical work, including field and time reports, and material requisitions.

Cable Splicer

Cable splicers install and repair single and multiple conductor insulated cables on utility poles and towers, as well as those buried underground or installed in underground conduits. They make various types of connections between electrical conductors, insulating the conductors and sealing cable joints with a lead sleeve wiped onto the lead sheathing of a cable assembly. The cable splicer must be familiar with the various methods and techniques employed in bonding and grounding lead sheaths of cables for all voltages. He or she must also be aware of the methods of weatherproofing and fireproofing cables. Additionally, cable splicers must be familiar with the various new kinds of splicing kits on the market that are used especially for insulating splices and newer types of cable that do not have a lead sheath.

Journeyman cable splicers' duties also include complicated construction and maintenance work on dead or energized underground distribution conductors and equipment. They install underground services and conduits, breakers and fittings, service boxes, meter boxes, current protective devices and conduits, breakers and fittings, service boxes, potential transformers, and motor and lighting circuits. The cable splicers' job also entails locating and repairing circuit and equipment faults; handling, forming, and bending all types of copper bus work; and mounting heavy equipment. Cable splicers must be familiar with the characteristics of metals and insulating materials used in their field of work. A knowledge of conduit and duct work is also essential.

Cable splicers work in manholes, vaults, and on poles when performing their duties, a substantial amount of which pertains to the installation and maintenance of underground lines and overhead cables and the changing of cable systems layouts. They also make all types of polyphase transformer installations, perform phasing and phase rotation tests on polyphase circuits, and, when required, trace and clear grounds and open circuits on cable systems. Occasionally, when work within this job classification is temporarily interrupted or delayed, the cable splicer may be assigned to other work within the occupational group.

The lineperson, troubleshooter, and cable splicer are, at times, required to work from detailed instructions and blueprints. They must be able to exercise judgment and independent thinking in arriving at good, workable decisions. The workers in these classifications must have patience, precision, manual dexterity, and the ability to exercise due care in connection with their work. Initiative and ingenuity are also essential attributes. Often these journeymen supervise other workers assigned to their work crew. Therefore, they must be familiar with the National Electric Safety Code which is a consensus code covering the construction, repair, and maintenance of overhead and underground electrical conductors and equipment.

The work of lineperson, troubleshooter, or cable splicer is mainly performed outdoors. Occasionally, they are required to work under adverse weather conditions, such as in the aftermath of a hurricane, storm, or tornado, in restoring electrical power to a community. Their duties demand the ability to climb, bend, and work in somewhat awkward positions. They should possess sufficient physical stamina to meet the requirements of their occupation and should also have good visual perception.

Journeymen employed on substation work normally provide themselves with the following tools: knife, hammer, pliers, hacksaw frame, 14-inch pipe wrench, large and small screwdrivers, combination square, 9-inch level, 8- and 10-inch adjustable end wrenches, chisel and punch set, toolbox, and a set of box-end wrenches. Those journeymen performing line work are usually required to furnish climbers, safety belts, body belts, pliers, hammers, wrench and hand connectors, wooden rules, screwdrivers, adjustable end wrenches, 9-inch lineperson's pliers, and wire-skinning knives. Line construction and maintenance is interesting and adventurous work. It offers a challenging career to those who want to contribute their knowledge and skills to a branch of the electrical industry that is vital to our nation's security and industrial growth.

ELECTRIC SIGN SERVICEPERSON

Every community, regardless of its size, has an attractive display of electric signs—from elementary signs advertising a neighborhood drugstore to the elaborate array of spectacular signs that line the main section of New York City's Broadway. Since their invention, luminous tube signs and illuminated plastic signs have been accepted as a valuable outdoor advertising medium. These signs advertise the names, products, and services of the hundreds of thousands of factories, stores, restaurants, hotels, theaters, and other business and commercial establishments across the country. They are installed, serviced, and repaired by electric sign service workers who are responsible for ensuring that the equipment they install and maintain operates virtually free of interruption. Hence, the electric sign serviceperson must be capable of discharging duties in an effective, proficient manner.

Installation

Electric sign crafts workers must have the ability to plan on-the-job layouts from detailed prints that illustrate construction design and materials used. They must be familiar with the proper methods of hanging signs. This, in itself, calls for substantial knowledge and skill in the art of sign suspension. Many signs are large and heavy and require special techniques to support them safely in forceful winds. Consequently, the sign journeyman must be cognizant of the various types of supports and their safe working loads. The sign erector should also have a good understanding of applied physics and should be thoroughly familiar with rigging.

Another area in which the sign serviceperson should have experience is metalworking. The worker must be able to bend sheet metal and plastic into various forms, cut out metal and plastic parts, and be familiar with drilling or punching holes, riveting, and soldering metalwork. It will also be to his or her advantage to be capable of performing operations involving acetylene burning and electric arc welding.

Repair

Service calls make up a large portion of the electric sign serviceperson's job. Whenever an electrical sign fails to operate properly, a worker is dispatched to remedy the failure. He or she must be able to diagnose the cause of the trouble expeditiously and to follow up with corrective action. To do this, the worker must have a comprehensive understanding of electric circuitry, as related specifically to electric signs, of transformer theory and application, of gases and their uses, and of mechanics, especially those pertaining to gears, drives, bearings, and other mechanical parts of revolving signs. Electric sign repair workers replace defective electrical wiring, transformers, flashers, time clocks, sockets, tubes and lamps, electric motors, and other mechanical devices and parts.

Electric sign journeymen should be acquainted with the National Electrical Code, particularly those sections relevant to the construction and maintenance of electrical signs. They should also be aware of local building code requirements which cover construction, erection, and maintenance of signs and outdoor display structures—with respect to safety, size, and attachment or anchorage.

Service

Another major portion of the electric sign serviceperson's duties comes under the category of preventive maintenance. Many firms purchase service contracts at the time they buy an advertising display; others purchase a sign maintenance contract after the initial purchase. In each case, the sign serviceperson performs periodic routine maintenance on the equipment covered by these maintenance contracts.

Sometimes sign service workers function as customer service personnel. That is, they occasionally suggest to customers ways to increase the attractiveness and visibility of advertising displays. For example, they may recommend changing the color scheme, attaching flashers, or raising the height of the display. Sometimes the worker is involved in selling service contracts to customers.

Preparation

Many sign journeymen invest up to $275 in tools. Their toolboxes contain basically the same hand tools as those used by the inside construction electrician. The test equipment used is normally furnished by the employer.

Electric sign journeymen must feel comfortable when working at various elevations. Often they must work from scaffolding, and they are, at times, required to use safety belts and lines. Because this type of work is performed almost exclusively in exterior locations, persons aspiring to careers in this field should be fond of working outdoors. They should also possess agility and the ability to climb well and be able to adjust readily to working at different heights and in various positions. The physical aspects of the job require medium strength, excellent sense of balance, and general good health.

The electric sign industry has good growth potential and should provide ample employment opportunities for those seeking careers.

CHAPTER 5

OPPORTUNITIES FOR ADVANCEMENT

Apprenticeship training provides an optimum training system for learning a skilled trade; it gives the trainee an opportunity to progress in an orderly manner from the entrance level to journeyman status. Once the trainee meets the criteria established for entrance into an occupation with apprenticeships, the progression is fairly well planned and assured—if the apprentice satisfies the requirements specified for advancement through the various levels of competency.

APPRENTICESHIP PROGRAMS

Apprenticeship training involves on-the-job training, supplemented by related classroom instruction. Most apprenticeship programs cover a five-year period. The five years are usually divided into ten equal intervals, each six months long. Normally, apprentices receive a set percentage of the journeyman's wage rate for each six months of training they complete. The following is an example of an apprentice wage schedule typical of those found in collective bargaining agreements.

First six months—35% of journeyman's wage rate
Second six months—40% of journeyman's wage rate
Third six months—45% of journeyman's wage rate
Fourth six months—50% of journeyman's wage rate
Fifth six months—55% of journeyman's wage rate
Sixth six months—60% of journeyman's wage rate
Seventh six months—65% of journeyman's wage rate
Eighth six months—70% of journeyman's wage rate
Ninth six months—75% of journeyman's wage rate
Tenth six months—85% of journeyman's wage rate

Most apprenticeship programs are developed along functional lines, not only to promote the growth of the apprentice's capabilities, but also to meet the ever-changing needs of our economy. Without this kind of organization, it is easy to overlook a training program's primary objectives. Some of the main goals for the development of well-rounded journeymen are pride in craftsmanship, initiative, and ingenuity. These goals must be interwoven into the fabric of the training program.

Apprenticeship is a training system based on a written agreement between the apprentice and the local joint apprenticeship training committee. The apprenticeship training committee usually consists of six members. Three of these members represent the employers, and three represent the union. They are responsible for conducting and supervising the apprenticeship program at the local level. They test, select, and indenture apprentices and register them with the U.S. Department of Labor's Bureau of Apprenticeship and Training or with the state apprenticeship agency. Local joint apprenticeship training committees often employ a training director to assist them in carrying out their duties.

Apprenticeship training committees are responsible for monitoring and evaluating the variety and the quality of the apprentice's performance. They implement job rotation to ensure that the apprentice receives varied experiences with the latest materials, equipment, and construction processes.

Apprenticeship training programs have certain advantages over less formal training programs. This is especially true of programs that are national in scope, since they enhance the quality of the training through the use of standardized training materials.

Completing an apprenticeship gives workers recognized status and dignity. It also provides them a margin of job security, and often it may increase opportunities for promotion to supervisory positions.

Another benefit of participation in apprenticeship training is the VA monthly training assistance allowed to a veteran pursuing a full-time approved apprenticeship. Certain benefits can be applied to on-the-job training that meets specified requirements. The following table illustrates the VA assistance schedule.

Period of Training	No Dependents	One Dependent	Two Dependents**
First 6 months	$274	$307	$336
Second 6 months	205	239	267
Third 6 months	136	171	198
Fourth 6 months*	68	101	131

*and any succeeding 6-month period
**An additional $13 for each dependent in excess of two

ON-THE-JOB TRAINING

Aside from completing an apprenticeship training program, trainees may become skilled trades workers by acquiring skills and knowledge through on-the-job experiences. On-the-job training (OJT) is an informal type of training, and so is usually not structured as well as the formal apprenticeship training program. The trainee endeavors to learn the trade by performing a variety of tasks and by assisting other skilled workers in the performance of their duties. Quite often the trainee does not receive any additional off-the-job training to supplement his or her work experiences.

There is general agreement that this type of training has drawbacks. It usually does not come under the auspices of a training committee, nor does the trainee enter into a written agreement covering a specified period of training. The lack of these two elements in a training program can present several problems that must be resolved for a young worker to become a well-rounded craftsperson within the training period recommended for that particular occupation. If they do not receive guidance from a policy-making body, their training can lack supervision and direction. Often journeymen are under constant pressure to meet production demands, and so have little time to devote to the learner. And since on-the-job training is usually unregulated, the trainees do not receive the variety of work assignments essential for the development of a competent craftsperson. Therefore, the time required to advance from the entry level to journeyman status is usually greater than that required in formal training programs.

Unlike the apprenticeship training program, which provides for orderly progression and specified hourly wages for each level of competency attained, the wages paid to workers learning their trade through on-the-job training are normally determined by the individual's particular job classification and the type of duties performed.

If administered in the proper manner, on-the-job training can lead to successful skill development. Adequate job rotation, monetary incentive for advancement, and job security must be included in the overall objectives of an effective training system for skilled workers.

In some situations, on-the-job training offers workers in companies that do not have apprenticeship programs the opportunity to acquire a trade. It also appeals to persons who are not interested in school work, since generally, there is no related classroom instruction involved. Although this type of training has played a significant role in the training of skilled workers in some occupations, it has not been a very popular way of attaining journeyman status in the electrical field.

EVENING EXTENSION CLASSES

Many vocational schools and community colleges offer evening courses in addition to the regular, daytime vocational instruction. Evening courses are planned to meet the needs of young adult workers attempting to supplement their on-the-job experiences with related theoretical training. The students attending these programs have an adult point of view regarding the method of instruction and the contents of the curriculum. Attendance is voluntary and lasts only as long as the students feel they are receiving the instruction they desire. The success of evening extension classes, therefore, depends mainly on giving students material relevant to their needs in industry. The methods of teaching and the range of material covered should be as flexible as possible, in order to meet the educational aims of the class.

Most of the difficulties that arise in evening school programs can be overcome if the administrator in charge of continuing education has a good grasp of the programs under his or her control. Above all, the administrator should make certain that the instructional staff possesses the expertise necessary to impart the skills and knowledge that are essential to developing occupational competency.

Attending evening school classes can be a fruitful experience for students who are really serious about advancing in their chosen vocation. These classes help bridge the gap between the practical applications of the trade and the theoretical principles which, together, help form a better understanding of the mechanics and science involved. Students enrolled in extension courses have the advantage of drawing upon the experiences of the instructor. They can ask questions and participate in classroom discussions which can be beneficial to the entire class. Instructors can, at times, shed light on a complex problem that may have arisen on one of the student's jobs. And they often can smooth the transition from a purely theoretical statement to one that incorporates both theory and practicality.

Usually evening extension programs are two years in length, although a few courses may cover a slightly longer period. The contents of these courses normally range from the fundamental aspects of the trade to an introduction to the more intricate theory and science associated with the field. Many of these courses include laboratory work which allows students to become familiar with the theory and application of electrical test equipment and circuitry, as well as the principles underlying the experiment.

The rate at which a student progresses through the course depends largely on individual initiative and aptitude. In some schools, students are allowed to enter the extension program at intermediate points during the school term, rather than waiting until the subsequent school term. However, once an individual decides to pursue this type of vocational education, he or she should make every possible effort to attend classes regularly and to actively participate in class studies and exercises.

CORRESPONDENCE COURSES

Home study courses are another way an individual may acquire theoretical training to supplement on-the-job work experience. This type of independent study is widely known as correspondence education, and it has both advantages and disadvantages. Perhaps one of its more attractive features is the latitude students have in allocating their spare time for study. However, this unrestrained method of study can present serious problems for students who lack initiative and determination. People who enroll in correspondence study courses must discipline themselves to use their spare time prudently; they should adhere to a study pattern that will allow them to accomplish their vocational objectives within the recommended period of time.

Another desirable feature of this method of study is that students learn by what they do. That is, they rely primarily on teaching themselves. This educational technique often helps to instill greater self-confidence. Usually, the course materials are designed to stimulate the student and bring about the desired response. The students, within limits, proceed with their work in accordance with their own abilities, so their study assignments have a fair degree of flexibility. They may choose to spend more time on the subjects which give them the most difficulty and less time on the ones they are able to grasp readily.

Correspondence course study allows students a reasonable amount of freedom in selecting courses which pertain to their major area of interest. Hence, students can subscribe to courses that serve their immediate needs.

Most correspondence course programs offer continuous levels of instruction from basic through advanced courses. And, normally, it is the students' decision to determine just how far they want to go in the program, once they have completed their basic block of instruction. The following is a sample of some of the home study courses that should prove valuable to persons desiring careers in the electrical field: mathematics, drafting, blueprint reading, electricity, electronics, industrial electronics, construction, building maintenance, refrigeration and air conditioning, solar heating, and welding.

For many years the National Home Study Council has been the standard-setting agency for home study schools. A part of this council is its accrediting commission which establishes educational, ethical, and business standards; it examines and evaluates home study schools in terms of these standards and accredits those that qualify. For additional information on correspondence education, you may wish to write: Executive Secretary, Accrediting Commission, National Home Study Council, 1601 Eighteenth Street, N.W., Washington, D.C., 20009.

Again, it should be emphasized that the success of this type of study program depends mainly on the individual. He or she must be properly motivated, possess proper study habits, be capable of working alone, and be self-disciplined to meet the requirements dictated by this method of study.

SKILL IMPROVEMENT PROGRAMS

Skill improvement programs usually are geared to advanced training or training in a specialized area above the apprentice level, while the training programs previously described normally apply to training below the journeyman status. Journeyman training programs provide workers with an excellent way of obtaining instruction on current installation practices and techniques, as well as furthering their knowledge of the theory associated with their fields of study.

The technical knowledge gained through attending skill improvement courses often qualifies the participating workers for key and supervisory positions in their field. By concentrating on instruction in a particular phase of their occupation, the journeymen can develop expertise in that area. For example, they may decide that they would like to become proficient electric motor control specialists. To accomplish this aim, they would focus their attention on that type of training.

Skill improvement courses are usually offered by vocational-technical institutes. Quite often, these courses are sponsored by labor unions; sometimes they are backed by management, and occasionally they are conducted on a joint basis. Many skill improvement courses consist of 15 units of instruction, which can be covered in a semester consisting of one three-hour period of instruction each week for 15 weeks, with additional time allowed for tests. Upon completion of the various units of instruction, the journeyman is awarded a certificate which signifies that he or she has satisfied the requirements for that particular course.

The main objective of skill improvement programs is to develop additional skills and technical competence. Also of prime consideration is inspiring the craftsperson to continue the quest for knowledge of the trade and to strive to produce quality work.

Skill improvement courses also increase the journeyman's opportunities for promotion to a supervisory-level job. Many supervisors and others in high administrative positions in industry have come from the ranks of crafts workers.

CHAPTER 6

SPECIALIZING IN RELATED FIELDS

Crafts workers sometimes find it beneficial to specialize in one particular area of their field or in some field related to their occupation. There are times when journeymen can increase their earning power by electing to specialize. For example, a journeyman wireman electrician may decide to become a control specialist or an estimator, while a journeyman lineperson may choose to embark on a career as an electronics journeyman or a supervisor. The maintenance electrician may find it advantageous to focus his or her talents on the field of instrumentation. Of course, there are many more options available to the skilled mechanic, and a number of these specialties will be discussed in subsequent sections of this chapter.

However, first let us consider some of the reasons why journeymen sometimes decide to specialize. As stated previously, specialization will, at times, lead to greater income for the worker. This may be brought about by the combination of a higher rate of pay coupled with stable employment. In addition to earnings, such factors as interest and aptitude usually play a significant role in determining one's progress in a particular field. The general working conditions associated with certain specialties often attract individuals into those areas of employment. For example, the work of a control specialist may not be as strenuous as that of some other branches of the trade. The duties of the instrumentation electrician may be considered less physically demanding by some.

Most craft workers are proud of the knowledge and skills they possess and are easily motivated. Thus, they are often inspired to work for additional goals simply because such work gives them a feeling of self-satisfaction. Once a journeyman is successful and respected by peers, he or she will often decide to pursue specific segments of the trade in depth.

The best way to approach specialization is, first, to establish a sound base on which to build. This base is essential if the specialist is to impart technical wisdom to other workers. Too often, costly problems on jobs arise because of inadequate communication between personnel. So it is fundamental that specialization follow, rather than precede, the development of skilled mechanics.

Persons who want to specialize should try to prepare themselves by studying materials relevant to their area of interest. Also, they should be ready to accept whatever challenges lie within the scope of their abilities. This provides them with the opportunity to apply their additional skills and gives them added confidence. Like the journeyman, the specialist must continue efforts to stay abreast of the field.

CONTROL SPECIALIST

Control specialists service and repair control equipment and control devices used for starting, accelerating, speed-regulating, braking, and stopping electrical machinery. In earlier years, this type of equipment was serviced by the journeyman electrician. However, increased technological sophistication in this field resulted in the need for specialists who possess expertise in the art of installing, servicing, and repairing intricate control systems. The specialist must not only be familiar with electrical theory and its application, but must also possess knowledge and experience in such fields as electronics, pneumatics, hydraulics, and mechanics.

Today's complex industrial production techniques, often fully automated processes, require knowledgeable personnel capable of minimizing and rectifying control problems, thus reducing costly stoppages of production.

Control specialists utilize a variety of test equipment to help them perform their jobs. This equipment includes voltmeters, ammeters, wattmeters, oscilloscopes, signal generators, capacitor checkers, resistance bridges, tachometers, frequency meters, and logic probes. The test equipment used by control specialists often enables them to locate faults that would be virtually impossible to pinpoint without such elaborate test instruments. Of course, they must also rely heavily on sight, smell, and touch. These senses can be very valuable to the control specialist when performing preliminary troubleshooting procedures.

Control specialists must be expert troubleshooters. Hence, they should be thoroughly familiar with all aspects of their field. Some of the devices and apparatus they are responsible for servicing include magnetic amplifiers, saturable reactors, magnetic clutches, amplidynes, Regulex and Rototrol rotary amplifiers, magnetic blowouts, linear and analog amplifiers, microprocessors, computer modules, and a host of other devices used in the control of electrical equipment and machinery. They must also be acquainted with the circuitry in which these devices are utilized.

A major portion of control specialists' work is performed indoors. Consequently, they usually work under favorable environmental conditions. Although some control specialists choose to own special tools such as hollow head nut drivers, small socket sets, bristle brushes, and special screwdrivers, their tools are basically the same as those used by the inside construction electrician.

Control specialists' employment is fairly stable, since they are often key employees and thus enjoy a degree of job security. Because they possess knowledge and skills beyond those required for journeyman status, they frequently receive additional compensation in the form of premium pay. This may range from 50 cents to more than two dollars an hour above the regular journeyman rate.

ELECTRICAL SERVICEPERSON

The duties of the electrical serviceperson vary from the simple task of replacing blown fuses to the more complex functions of troubleshooting and repairing sophisticated electrical machinery and equipment. Most service workers are assigned trucks which are equipped to handle service calls. Some service workers are radio-dispatched, while others receive their assignments on a daily basis. Like the control specialist, the serviceperson must have a good background. It is essential, therefore, that these specialists be drawn from the ranks of experienced mechanics.

Electric motorshop journeymen and inside construction journeymen who enjoy the challenge of diagnosing circuit and machinery faults usually perform well as service workers. This specialty is also attractive to other journeymen in the electrical field, especially those versed in the art of troubleshooting.

Service workers usually work alone. However, there are times when another mechanic or apprentice will accompany and assist them in handling service calls that require additional workers. The time they spend on assigned jobs depends on the difficulty they have resolving the problems. Normally, they carry a variety of test equipment in their truck. They also rely heavily on their senses as invaluable troubleshooting aids.

A major part of serviceperson's work is done indoors. Their work is fairly clean and interesting and requires only moderate physical strength. Because of their special skills, they usually enjoy stable employment, and occasionally they have the opportunity to supplement their earnings by working overtime.

Service workers are sometimes permitted to use the company's service truck as their means of transportation to and from work. Their toolboxes contain essentially the same tools used by inside construction journeymen. However, many service workers add special hand tools to their kits to help them make repairs that require specific tools.

Electrical service workers are exposed to a variety of work assignments, each of which offers a slightly different challenge, so they are seldom bored. Also, they frequently derive pleasure from the fact that they have been able to resolve difficult electrical problems that have meant inconvenience, costly shutdowns, or both.

Since their daily work frequently places additional responsibilities on them, and because they provide their own supervision, they generally receive a premium of one to three dollars an hour above the journeyman wage.

ELECTRONICS JOURNEYMAN

Over the past couple of decades, electronics has become more closely woven into the field of electricity. At one time, there was very little need for an electrician to have a knowledge of electronics. This is no longer the case. In fact, most electrical trade curriculums presently incorporate some electronics training. This training serves not only as an effective base upon which one can easily build, but also as a stimulus for inspiring further study of the subject.

Usually those workers who install, service, and repair electrical-electronic equipment enjoy favorable working conditions. This is another incentive for becoming more proficient. Some electricians enroll in night study classes to accomplish this task, while others enter individualized training programs, such as correspondence course study.

The electronics journeyman's duties are diversified. He or she must be capable of performing a variety of tasks—from the simple soldering of a wire to a terminal to the more complex assignments, such as diagnosing troubles in sophisticated electronic equipment. Electronics journeymen must be adept in the art of troubleshooting, since a substantial part of their work relates to locating and correcting circuit faults in electrical-electronic equipment. Often, a piece of malfunctioning electronic equipment can disrupt an entire assembly-line operation, and unless this equipment is repaired efficiently and put back in service quickly, costly shutdowns can result. Some of the electronic equipment serviced by electronics journeymen includes electronic precipitators, induction and dielectric heating apparatus, electronic speed control devices for electric motors, various electronic sensing devices used for determining the level and/or temperature of a material, electronic controls used in conjunction with heating and air conditioning equipment, electronic power converters, computers, and communications systems. This is only a sampling of the electronic devices and equipment the electronics journeyman is responsible for installing, servicing, and repairing. Some electronics journeymen further specialize in certain areas, such as communications, computer systems, or entertainment devices.

In addition to the normal test equipment and instruments journeyman electricians use, they must be familiar with such test equipment as oscilloscopes, electronic voltmeters, various types of signal generators, power supplies, electronic switching devices, electronic tube and transistor testers, and capacitor testers. Expertise in the use of this test equipment allows the journeyman to pinpoint troubles swiftly in equipment that, without its use, might not be readily detected. The electronics journeyman's tool kit usually consists of a variety of small hand tools. The toolbox also contains a number of special-purpose tools designed to perform specific functions.

Electronics journeymen primarily work indoors, and their work does not require great physical strength. However, they should be alert and should have the attributes of a good troubleshooter. Like the control specialist, they customarily receive a premium rate of pay.

The career of an electronics journeyman can be very rewarding and exciting, especially for the person who enjoys challenges.

INSTRUMENTATION TECHNICIAN

Another specialty within the scope of the journeyman electrician's work is the installation, service, and repair of intricate instrumentation and control systems. These systems are employed to monitor, measure, record, and regulate production operations automatically. This field is becoming increasingly attractive to the journeyman wireperson, simply because he or she receiving more exposure to this type of work on today's industrial jobs. As our technology continues to expand, even more opportunities will become available for the highly skilled electrician who possesses sufficient technical competence in this field. These workers install, service, and repair instruments and control devices used to process chemicals; refine oil; control the movement of ships, aircraft, and missiles; generate electricity; and manufacture steel and a host of other products.

Quite often, sophisticated sensory systems employ a combination of devices which are sensitive to more than one type of energy source. Obviously, industrial instrumentation specialists must be familiar with the disciplines of electricity, electronics, pneumatics, and hydraulics if they intend to successfully meet the challenges offered by this field.

In addition to having an inclination for mechanics, the instrumentation specialist should also have a good grasp of technical skills associated with this type of work. Deftness is invaluable to instrumentation specialists. They are frequently required to work with small components in confined spaces, and many of the minute adjustments and close calibrations they must make require a very minor tolerance. They should also have the ability to interpret, read, and work from complex circuit drawings.

A significant part of instrument mechanics' work is performed under shelter. However, there are times when the job calls for servicing equipment that is outdoors and in remote locations. Nevertheless, they usually have fairly stable employment and good earnings, and, because of the additional skills and responsibility demanded by this type of employment, they often receive higher hourly wages.

Like the control specialist and the electronics journeyman, instrumentation technicians utilize a wide variety of test instruments and equipment in troubleshooting, and many of their tools also are quite similar.

Good sight, acute hearing, and a keen sense of smell are essential physical characteristics for instrumentation technicians. Usually, their work is not strenuous and does not demand great physical strength; however, it does require good finger dexterity.

Instrumentation technology is intimately woven into the basic fabric of our nation's technological growth. Thus, as our industrial technology expands, so does this branch of the electrical industry. Opportunities for those seeking a career in this specialty are virtually unlimited. Nearly all of our industrial plants utilize systems incorporating process controls, which are serviced by the instrumentation technician.

COMMUNICATIONS MECHANIC

The successful transmission of intelligence over telegraph lines using electrical impulses coded to represent letters and numbers was first demonstrated by Samuel Morse in 1835. Out of this humble beginning grew today's massive communications industry, which encompasses such systems as telephone, radiotelephone, computers, microwave, television, and others. In addition, there are in use such specialized systems as radar, sonar, telemetering, and loran which depend on the transmission and reception of electro-magnetic waves. The installation, service, and repair of this equipment and its associated circuitry is the responsibility of communications mechanics. Some of their other duties include installing, maintaining, and repairing mobile and fixed station radio transmitters and receivers; performing necessary adjustments on microwave equipment; servicing antenna and wave guide systems; analyzing and developing solutions for the various problems that occur in electronic equipment; responding to trouble calls on communications lines; and occasionally climbing poles to make temporary or permanent repairs on lines.

Communications mechanics must be adept in the use of test instruments and equipment. They often use circuit analyzers; audio, carrier, and radio frequency oscillators; carrier frequency voltmeters; oscilloscopes; and other test equipment. Communications mechanics isolate trouble and determine signal losses, gains, and levels. They also locate short circuits and defective circuit components such as defective tubes, semiconductors, integrated circuits, transformers, resistors, condensors, and other electronic components. They must be able to read and draw electrical and electronic circuit diagrams. Basically, they use the same hand tools as the electronics journeyman.

Communications mechanics should have good judgment and ingenuity. They must be able to plan, lay out, organize, and supervise, as well as performing nonroutine work.

Employment opportunities in this branch of the electrical industry are expected to grow rather slowly throughout the 1990s. Although demand for the services of the communications industry will continue to increase, advances in technology are expected to dampen employment growth. Technological changes in telephone communications should be particularly significant, with computers and other sophisticated electronic equipment playing a greater role in performing automatic functions, thereby reducing manual operations. Nevertheless, there will be a substantial number of job openings for those capable of mastering the intricacies of this industry.

Communications mechanics perform their jobs in a variety of surroundings. One day they may be working on a tower, installing a wave guide system for a radar unit. The following day they may be making adjustments on a transmitter in confined quarters. Communications work never ceases to challenge even the most talented troubleshooters, nor is it lacking in interest or excitement.

The work of communications mechanics does not demand great physical effort, so medium strength suffices for most of their duties. They should be in sound physical condition and should possess good senses of sight, smell, and hearing.

SALES OR SERVICE REPRESENTATIVE

Sales or service positions provide interesting and rewarding careers for a substantial number of workers in the electrical utility industry. The skills offered by these workers are essential for the effective operation of an electric power company. In addition to being technically competent, sales and service representatives must practice diplomacy. Frequently the job responsibilities require either direct personal contact or telephone contact with architects, general contractors, electrical contractors, builders, realtors, and customers.

Some of the specific duties of sales and service representatives are to promote the sale of residential, commercial, and industrial electric service; negotiate and attempt to secure contracts from customers for the extension of the company's distribution system; assist company and dealer salespersons in selling appliances, adequate wiring or other electrical installations; contact customers pursuant to inquiries and complaints regarding bills, rates, or services; verify the accuracy of meter readings, conduct investigations, and advise customers about the economical use of their electrical equipment; arrange for the installation of electrical services and meters; prepare sketches and various types of work orders, material lists, and other reports in connection with new business or the modification of existing systems; and, in some instances, be sufficiently qualified as a lineperson to perform certain emergency repair work. The scope of their duties subjects them to both indoor and outdoor assignments.

Customer service personnel should be thoroughly familiar with company policies, rules and regulations, rate schedules, and territory serviced. Their personal traits should include the ability to speak clearly and a temperament that reflects courtesy and tact.

SUPERVISOR

Thus far, we have focused our discussion on several trade specialties. Ordinarily, specialization follows journeymanship in a related area and thus affords those who gain additional knowledge and skills the opportunity for advancement. But this section discusses another way in which one may develop his or her career. Those who have attained superior knowledge and skills within their area and have demonstrated leadership qualities are often offered supervisory positions. Each supervisory position in the electrical industry requires a particular set of skills and talents; however, they usually have one requirement in common—experience in their respective job areas.

Effective foremanship is predicated on effective leadership. The adroit foreman is not only a master of the trade, but also is a leader. Organizational ability, judgment, integrity, initiative, and interest are essential properties of leadership. Supervisory personnel must possess these traits in order to perform their duties successfully.

Supervisors are responsible for making decisions, giving instructions, evaluating the performance of workers, coordinating job functions, and carrying out administrative procedures related to the various records they maintain. Most importantly, supervisors must have concern for the safety and well-being of their co-workers. They must also have the ability to cooperate with crew members and to train and assist their subordinates. The degree to which they fulfill these obligations depends, in large measure, on how effectively they apply the basic principles of leadership.

Electrical construction supervisors, line crew chiefs, and shop supervisors serve as a connecting link between upper management and labor union representatives. They must fully understand the terms of the labor agreement under which the members of their crew work so that they can perform their duties in a consistent, fair manner. Besides being adept in communicating with their associates, supervisors must be able to converse intelligently with engineers, architects, electrical inspectors, safety personnel, and the public. They must be both technicians and diplomats.

The job of a supervisor in the electrical industry requires the ability to plan, implement, and direct the installation, service, and repair of electrical equipment and systems. Supervisors must be able to transform drawings, instructions, and related information into working systems. They should be familiar with local and national electrical codes and with safety practices applicable to their branch of the industry. Because supervisors are leaders as well as artisans of their trade, they receive an additional hourly compensation which may range from 50 cents per hour to more than three dollars per hour over the pay scale of the journeymen assigned to them.

In brief, supervisors have the authority and responsibility for seeing that jobs are brought to successful completion in a cost effective manner.

ESTIMATOR

Interesting and challenging careers await those who can measure up to the standards of the estimating profession. The talents of accomplished estimators in the electrical construction industry are always marketable at respectable salaries. Every firm in the electrical contracting business must employ someone who is capable of functioning as an estimator—whether he or she is the owner, a supervisor, or a skillfully trained specialist. In general, most companies find it more efficient and more profitable to entrust the estimating of contract costs to a specialist charged with that responsibility alone.

Good estimating requires a combination of engineering knowledge, skill, and practical experience—plus facility with the tools and techniques of estimating. Electrical estimators must be familiar with the various tables, charts, and graphs used repeatedly in their field of work. The ability to interpret blueprints and other drawings is a basic necessity for this specialty. Other personal characteristics essential to the estimator are an acute eye and prudent judgment. They must be able to anticipate, within reasonable limits, on-site construction changes which could substantially alter the cost of a job. Estimates should be accurate, and appraisals of construction and installation costs factual. The estimator must approximate data as closely as possible and present it clearly and methodically. Personal judgments and guesswork are unavoidable, but they should be limited strictly to those areas where actual values or conditions cannot be predicted.

Estimators work mostly indoors in an office atmosphere. The job is not characterized by strenuous activity. However, there are times when estimators must work at a pace which can become quite demanding, especially when they are required to estimate contract costs under accelerated conditions. Electrical estimators normally have the same workweek as other office personnel, but occasionally their hours of work will differ according to the volume of work they are assigned and the time they are allowed to complete their assignments. Successful estimators enjoy steady employment, good working conditions, and respectable salaries. Frequently, they also share in the profits generated by the firm.

ELECTRICAL INSPECTOR

Electrical inspectors render invaluable services to municipalities, to the electrical construction industry, and to the consumer. Because of the hazardous nature of electricity, many precautions must be taken when installing electrical equipment and systems in residences and other structures. By following the safe practices set forth by various building and electrical codes, hazard-free installations are accomplished. It is the electrical inspector's job to see that this condition is fulfilled. Thus, inspectors are entrusted with the responsibility of determining whether or not electrical installations conform with acceptable standards. Inspectors are usually guided by written codes such as the National Electrical Code or any other rules and regulations governing electrical work that have been adopted by a regulatory body having jurisdiction within a particular community or defined area.

Electrical inspectors must be knowledgeable in both the technical and practical phases of the electrical trade. They should have intimate knowledge of the National Electrical Code, as well as any other rules and regulations that apply to their work. Inspectors should be able to communicate effectively with electric power company personnel, electrical contractors and their representatives, and the general public. They are required to submit reports, write letters, and keep records pertinent to their field assignments; hence, a substantial portion of their duties is administrative in nature.

Sometimes electrical inspectors are required to conduct inspections in places that are obscure, inconvenient, or both. Thus, they draw heavily on such personal characteristics as good vision, alertness, and agility to aid them in carrying out physical inspections of electrical installations. Inspectors should be mindful of details because what might, on the surface, appear to be a trivial discrepancy could, if overlooked, result in a catastrophe. Therefore they must always be conscious of the responsibility entrusted to them.

Most electrical inspectors are employed by municipalities, but some are employed by states. They enjoy steady employment and favorable working conditions, and many times their salary is based on a monthly pay schedule, with the medium range between $1,800 to $2,900 per month. Chief electrical inspectors naturally receive higher earnings. Since the job ordinarily requires the use of an automobile, transportation is furnished or an allowance is given for the use of a private vehicle.

In addition to the satisfaction derived from performing their duties, skilled inspectors enjoy the reputation and status of being experts in their field.

ELECTRICAL CONTRACTOR

Electrical contracting can be a lucrative and promising field for those who possess entrepreneurial ability as well as a knowledge of the trade. While most of the specialties discussed so far have focused heavily on the acquisition of additional trade and technical skills, the art of electrical contracting relies fundamentally on the unique combination of business management and technical skills. The experience of contractors and their key employees often makes the difference between success and failure in the electrical contracting industry.

Many a contractor started out at the grass roots level "working with tools" as an apprentice or journeyman. In a number of cases, they have also put in some time as estimators, supervisors, or operations managers. Such practical experience often makes decisions easier when it comes to planning, scheduling, and controlling business operations. But because the electrical contracting industry is becoming more complex, education and business experience are increasingly important in determining the dollar volume of the business in which contractors are likely to be engaged. A survey compiled by the National Electrical Contractors Association (NECA) shows that nearly half of the biggest contractors in the country are college graduates. More than two-thirds of these graduates earned degrees in engineering, with majors in business administration the runner-up. In short, success as a contractor depends on one's education and experience, plus initiative, foresight, judgment, personality, and the ability to communicate.

ENGINEER

Engineering is another area of employment within the reach of many of today's journeymen. High standards of apprenticeship training, coupled with the rise in the general education level of the United States population, provide an excellent base and an incentive for one to further his or her scholastic achievement. Also, the availability of vocational-technical schools and scholarship programs and an increasing number of junior colleges and four-year schools bring technical and professional education within the grasp of more people.

Engineers and technicians usually concentrate their technical knowledge and skills in specific areas, such as electrical power, electronics, electrical equipment manufacturing, communications, electrical maintenance, or engineering consulting. This specialization has stemmed, in part, from our country's technological growth, and from the development of automated equipment and intricate electrical systems.

Those who want to be engineers or technicians should have an aptitude for mathematics, physics, chemistry, engineering drawing, and mechanics. They must be able to translate engineering concepts into realistic programs and have the tenacity to stick with an assignment and carry it through to completion, no matter how demanding or intricate it may become.

TEACHER

Many vocational-industrial schools, technical institutes, and junior colleges recruit teachers from the country's industrial communities. In their selection procedure, the recruiting programs take into account professional training, experience, educational background, technical knowledge, professional attitude, and personality. In selecting personnel for instructorships in either trade or technician training, emphasis is usually placed on the individual's professional background rather than on academic accomplishments. So there are many teaching positions available for craftspeople and other skilled technicians who are capable of imparting knowledge to and cultivating skills in others.

Instructors must understand their field thoroughly and know how to teach it to others. Moreover, they should be able to gear their instructional techniques to reach slow and average, as well as superior, students. Teaching careers can be fruitful and rewarding, especially for those engineers, technicians, and artisans who enjoy assisting and working with other people.

CHAPTER 7

EMPLOYMENT AND EARNINGS

Traditionally, the earnings of workers in the electrical industry are as good as or better than the earnings of workers in similar industries. For example, electricians are among the highest-paid workers in the construction industry. Also, there has been steady growth in the earnings of utility workers, maintenance electricians, shop journeymen, marine electricians, and electric sign service workers. The annual earnings of workers in these branches of the electrical industry usually provide sufficient income to maintain a comfortable standard of living.

WAGES

Construction electricians, like most of the other classifications discussed in this book, receive hourly wages. Although the terms *wages* and *earnings* are often used interchangeably, they actually differ in precise meaning. Wage rates stipulate a specific amount of money for a stated period of time (normally an hour) whereas earnings refer to the total amount of compensation received and are affected by the number of hours worked, overtime premiums, holiday pay, and other factors.

Wages and earnings depend upon several factors: (1) the type of work being performed, (2) whether or not the worker is covered by a union contract, and (3) the geographical section of the country where the work is being performed. The accompanying table illustrates the average wage rate for several classifications which have been discussed previously.

TABLE 1 Average Hourly Wage Rates for Journeymen
(Spring 1988)

Job Classification	Industry	Average Wage Rate
Construction Electrician	Inside Construction	$17.72
Sign Serviceperson	Electric Sign	$13.91
Lineperson	Private Utility	$15.88
Troubleshooter	Private Utility	$16.43
Maintenance Electrician	Electrical Maintenance	$14.87
Electric Motor Repairer	Electric Motor Shop	$13.20
Marine Electrician	Ship Building & Repair	$12.82

FRINGE BENEFITS

A typical contract in the electrical industry provides for substantial fringe benefits, and these benefits play an important part in contract negotiations. As a result, electrical workers are enjoying longer vacations, more paid holidays, better pension and insurance plans, additional sick and bereavement leave, and such other benefits as supplemental unemployment compensation and severance pay allowances.

Vacation plans for construction electricians are normally financed on the cents-per-hour-worked basis described in chapter 1. Vacation schedules which provide for vacations of one week after six months service, two weeks after one year, three weeks after five years, and four weeks after fifteen years of service are common in some branches of the electrical industry.

The number of paid holidays continues to grow. A recent survey indicated that a large percentage of labor contracts now provide for ten or more paid holidays a year. The same study showed an increase in the number of workers receiving 11 or more paid holidays per year. Also, many workers in the electrical industry receive premium pay for holidays worked.

Supplemental unemployment benefits (SUB) and severance pay allowances are gaining in popularity. These two benefits give the worker a measure of security during such crucial periods as layoffs and terminations. A large number of electrical workers are covered by some type of SUB or layoff benefit plan, and many are covered by some form of severance pay plan.

HOURS OF WORK

Electrical workers' workdays and workweeks mainly depend on the branch of the industry in which they are employed. Construction electricians usually work straight days, while many workers in other branches of the industry work on a shift basis. Straight day work may be defined as the period of work from approximately 8 A.M. to 4:30 P.M., Monday through Friday, with a half-hour lunch period.

A substantial number of labor agreements provide for workweeks of fewer than 40 hours, and the number of such contracts continues to grow each year as more and more workers express a desire for longer periods of leisure time. The International Brotherhood of Electrical Workers has over 700 collective bargaining agreements with employers in the electrical industry providing for workweeks of fewer than 40 hours. These contracts are found in such branches of the industry as construction, maintenance, communications, utility, repair and service, and others.

LOCATION OF EMPLOYMENT

The location of employment for electrical workers is determined by the branch of the industry in which they are employed. For some sectors of the trade, the job site remains relatively constant, while in others the place of work changes and is governed by building and construction activity. Both the geography of an area and its natural resources have a bearing on the type of employment likely to be found there.

Highly industrialized areas provide substantial employment for maintenance electricians, electric motor repair workers, utility and power plant workers, and construction workers. An electric sign serviceperson can find good employment opportunities in large commercial areas where businesses have huge displays in illuminated plastic and metal fixtures to advertise their products and services. Port cities—cities having navigable waterways and shipyard facilities—offer employment to marine electricians.

In summary, the type of industry, the available resources, and the natural features of the land influence the types of employment associated with a given locality.

EMPLOYMENT SECURITY

The degree of employment security one has in the electrical industry depends on several conditions. The state of our nation's economy, weather conditions, and other seasonal factors each have a bearing on the level and stability of employment. The electrical industry, like other industries, is affected by each of these elements, and some branches are influenced more than others by particular conditions. Moreover, while general economic conditions are likely to affect employment in all branches of the electrical industry, some segments will experience less fluctuation than others.

Employment in the construction industry is more affected by seasonal conditions than most of the other branches of the industry. The lack of building and construction work has been a continuous problem for the construction worker and has contributed to the underutilization of construction resources—especially human resources. To alleviate seasonal fluctuations, the federal government has adopted a counterseasonal policy—a program designed to help strike a balance between the lack of construction activity during the winter season and the peak activity of the summer season. Fortunately, employment for the inside construction electrician is not quite as unstable as for some of the other building trades, since a substantial amount of the work is performed under cover.

Normally, maintenance electricians and electric motor shop repairers work on a steady basis and are seldom required to take time off without pay. Generally, their work is unaffected by inclement weather conditions, and even during business lulls their employers are usually able to keep them gainfully employed in repairing idle machinery and its associated equipment or in servicing the equipment they use in carrying out their job functions.

Seasonal changes have little influence on marine electrician's employment. But work in this field is governed, to some extent, by the amount of shipbuilding taking place. And shipbuilding, like other construction work, lacks constancy. But since many marine electricians also seek employment in the repair and service specialty of the marine field, a large number of them enjoy reasonably stable employment.

Workers in the electric utility industry usually have full-time work. While the number of hours they work per week may vary, from a high in the peak summer periods to a low in less power-demanding winter months, they usually have continued employment. Linemen and other utility workers will average workweeks of 40 hours or more.

Because more of the electric sign serviceperson's work is performed out of doors, weather conditions play an important role in determining whether or not a job may be put in place. Consequently, overall conditions of employment for electric sign workers are influenced by both general business conditions and the weather. But employment in this field is reasonably steady.

JOB OPENINGS

By 2000 our country's population is projected to increase to around 268 million, and during that same period the nation's economy is expected to experience an annual real growth rate of 2.4 percent. Population growth spurs the need for workers to provide more housing, medical care facilities, educational institutions, and other services and goods. In a similar fashion, as our nation's economy grows, it generates employment opportunities. Obviously, not all segments of the American economy expand at the same growth rate. For example, the service-producing industries are expected to experience the highest percentage of growth through the 1990s and on into the next century. One of the industrial sectors that falls within this group is public utilities, and effort has been made in this book to describe the electrical jobs that can be found in this industrial sector.

The consumption of electrical power has declined over the last decade resulting in diminishing need for electrical power companies to increase their generating plant capacities. However, many utility companies throughout the United States have spent considerable capital expanding and upgrading their transmission and distribution systems. And, although the overall growth rate of electricity has slowed, there are some regions of the country where power companies have had to expand their generating capacities. Many of these power companies will be continuing their expansion programs. Improved energy efficiency and greater electrification will continue to improve our nation's productivity and competitiveness.

The capital expenditures required for modernization and growth in the utility industry are enormous. The construction of electric power plants and substations provides many good job opportunities. Generally, as utility companies increase their generating capability, they must also increase the number and capacity of their transmission and distribution lines to carry the added output. Understandably, this results in more jobs for construction workers and line crews.

TABLE 2 Employment Data
(Projections)

Occupation	1988*	2000*	Increase
All Construction Trades	3,807	4,423	16%
Electricians	542	638	18%
Communication Equipment Mechanics Installers and Repairers	113	94	−16%
Electrical and Electronic Equipment Mechanics, Installers, and Repairers	533	586	10%
Electrical Powerline Installers and Repairers	104	122	17%
Electronics Repairers—Commercial and Industrial Equipment	79	92	17%

(Source: U.S. Department of Labor, Bureau of Labor Statistics)
*Employment in thousands; ex., in 1988, there were 542,000 electricians.

Goods-producing industries, of which contract construction is a part, are expected to increase in employment by about 16 percent over the period 1988–2000. Despite this considerable employment growth, goods-producing industries will not increase as rapidly as the service-producing industries. However, for the sector of contract construction, employment is likely to increase significantly over the next few years as our nation turns to rebuilding our infrastructure and strives to meet the goal of adequate housing for our citizenry. The Bureau of Labor Statistics expects employment in contract construction to rise from about 5.1 million to about 5.8 million by 2000.

For crafts workers, as a group, employment is projected to reach a little over 15 million—nearly a 10 percent increase by 2000. As more industrial plants, commercial establishments, residential, and public facilities are built, the need for more workers in the construction market is created. Numerous job openings are also created for workers to service and maintain these new facilities.

The need for more workers provides only a part of the picture of long-term employer needs. A substantial number of job openings result from retirements, deaths, occupational shifts, upgrading and specialization, and other labor force separations. Thus, the total number of job openings for a particular occupation depends both on industry growth and work force replacements.

Detailed employment growth data for electric motor shop journeymen, marine electricians, and utility and power plant occupations is not published, but employment in each of these fields is expected to grow moderately over the next decade. Opportunities for workers in these occupations should increase steadily as the nation's economy expands—creating a need for an additional 18 million workers. This will bring the total work force to about 136 million by the year 2000.

TOMORROW'S JOBS

Through the 1990s the job structure for skilled workers in the electrical industry will not change drastically. Nonetheless, the contents of most major job classifications within the industry will experience some change as new products, equipment, and machinery emerge. New installation methods and other technological accomplishments will also have an influence on industry employment and occupational requirements.

The increasing use of prefabricated components and modular construction techniques will undoubtedly continue through the 1990s, especially in the areas of preassembled industrial equipment and residential construction. The net result will be more hours of work performed in manufacturing plants and fewer hours spent at the job site.

Solving environmental problems, advancing space exploration and travel, alleviating transportation problems, and eliminating the uneconomical use of our natural resources will each have a major role in upgrading our standard of living. Activity in these areas will result in the creation of new jobs and in many changes in the structure of existing jobs. Because much of the technological progress of our nation is the result of our success in electricity and electronics, these industries will probably grow considerably with new discoveries and inventions.

Energy Resources

During the 1990s, our government must develop comprehensive policies and programs to enable our country to become self-sufficient in terms of our energy requirements. With the domestic supply of oil diminishing, there is a need for the nation to develop alternative sources of energy. Also, it will undoubtedly become necessary for us to examine our present use of various energy sources so that we may assign priorities to their use. Hence, it is likely that we will see a shift toward a greater use of electrical power for such things as heating homes, businesses, and public buildings. Both baseboard resistance element heating and the heat pump can play important roles in meeting this objective. This shift in the use of our energy resources would allow us to conserve oil reserves for uses which petroleum is uniquely qualified to serve.

In the 1970s, the United States aimed at making nuclear power an important component in the total mix of sources used to generate electricity. However, both the increasing costs of constructing nuclear power plants and the safety concerns have limited the use of nuclear-generated power. Nonetheless, it is essential that our government, with input from appropriate advisory groups, develop a long-term energy policy with specific time-targeted goals. For the United States to remain competitive, reliable and affordable electrical energy will need to be produced. As we address our national energy needs, and as electrical energy fills an increasing proportion of those needs, more jobs will become available in the electrical industry.

The Environment

The fight for clean land and air and unpolluted water will require long-range planning and prudent policy decisions by our country's leaders. The success of our endeavors depends, largely, on (1) the development of new and better electronic precipitators capable of removing larger quantities of pollutants from the atmosphere, (2) more elaborate water purification systems which will utilize electrical and electronic components, and (3) more sophisticated electrified sanitary centers to deal with the disposal of solid wastes.

Another problem in many of our major cities is outmoded, inadequate transportation systems. Consequently, millions of automobiles, each carrying about 1.3 persons, are used each day to transport people to and from work. This compounds our transporation problems and adds to our environmental problems, since these automobiles are responsible for the emission of thousands of tons of pollutants into the atmosphere daily.

Urban Renewal

During the 1990s substantial progress should be made toward eliminating many problems associated with large city transportation systems. A number of cities have either begun installing, or plan to install, modern transportation facilities. The rapid transportation systems presently in operation are highly dependent on the use of electricity. They utilize electrical energy for propulsion and involve elaborate electrical wiring networks and intricate control systems which incorporate computers and microprocessors. Thus, the technology of electricity and electronics will provide the primary thrust for the development of modern, efficient rapid transit systems. Undoubtedly, this area holds significant potential for alleviating our transportation problems and providing substantial job opportunities.

There is a great need to revitalize many of our nation's cities. Such a program is absolutely necessary not only from a human point of view, but also from an economic perspective. Much of this country's industrial production takes place in large cities that are in need of rebuilding. Commercial establishments, as well, are in need of repair and replenishment. Housing is another area that certainly needs to be addressed in the 1990s.

Such an urban renewal program would entail new construction work and modernization and/or repair of existing buildings. Needless to say, this would generate substantial employment opportunities for crafts workers as well as numerous other occupations.

It would also add significantly to the economic well-being of our nation.

New Technologies

Space exploration and travel should continue to provide new employment opportunities, particularly the space shuttle program as it moves forward. Many electrical workers are presently employed in this sector. Some of the more populated occupations are electricians, electronics journeymen, instrumentation mechanics, test technicians, and control specialists.

Two new emerging technologies that are likely to have substantial impact on the electrical-electronics industry are fiber optics and robotics. The use of fiber optics has tremendous potential in the telecommunications industry. In 1966, scientists working for International Telephone and Telegraph (ITT) made lightwave communication practical. Four years later Corning Glass Works provided the main ingredient for a working system with development of an optically perfect glass fiber that solved the earlier transmission loss problems. Today, fiber optic cable systems are being installed in increasing numbers.

The development of HDTV (High Definition TV) is sure to bring many innovations to those electronic systems that employ basically the same visual technology the television industry presently uses. The considerable improvements in visual quality that will come forth as HDTV is developed are likely to improve various systems used by the armed forces of the United States. This technology could also result in major improvements in the field of radiology. The nation's development of HDTV technology holds significant promise for our production of consumer electronic goods, which is vital to our microelectronics industry. Advancement of this technology should result in many job opportunities.

Computer technology is also playing an important role in the communications industry. Electronic switching enables telecommunications companies to expand their business into a host of new markets. Computers and their associated software programs allow companies to transmit all types of data across the nation and around the world.

Robotics—the use of computers with machine-tool equipment—allows manufacturers to produce relatively small batches of products using programmable automation. For a number of years industry has employed fixed or hard automation in its production processes. While this type of automation lends itself very well to standardized types of products, it is not effective or economical when products are manufactured in smaller quantities. However, as it reaches maturity, robotic technology should offer a solution to the shortcomings of hard automation.

Since robotic technology incorporates the use of computer systems, sophisticated machine-tool equipment, and intricate wiring and control systems, considerable work opportunities should arise for those who are able to produce, install, service, and repair this equipment. This technology is expected to experience substantial growth in both development and application during the 1990s.

In summary, the 1990s should bring forth the opening of a considerable number of new employment opportunities in addition to the normal annual job openings that occur in the electrical industry.

CHAPTER 8

ORGANIZATIONS IN THE FIELD

The purpose of this chapter is to introduce some of the important organizations associated with the electrical industry. Each has its own particular objectives and functions, but each contributes to the success and progress of the industry. Because of the large number of organizations connected with the electrical field, we will limit our discussion to those that relate closely to the theme of this book.

ORGANIZED LABOR

Organized labor represents approximately one-sixth of our nation's work force. It consists of those workers who have joined labor unions to advance their cause—the promotion of their economic and social interests. Although the early labor movement in this country had few of the characteristics of present labor unions, it did bring workers together to consider problems of mutual concern and to devise means for their solution. Also, it enabled workers to realize that unity gave them strength, and so it often allowed them to resist unfair employer demands, such as wage reductions and unreasonable hours of work. As early as 1778, workers had banded together in an effort to secure better wages, minimum wage rates, shorter hours, apprenticeship regulations for crafts, and the advancement of union labor. The formation of craft unions did not come easily or without stiff employer opposition and much government interference. Unions were prosecuted as "conspiracies in restraint of trade" under an old English common law doctrine. As courts attempted to apply this conspiracy doctrine, it became a very controversial issue. And even when the courts began to back off from the conspiracy issue, they began to focus their attention and restraints on the tools which unions had available for securing better wages and working conditions—the strike and the boycott.

National Unions

Because of the constant opposition unions faced and the downward sweeps in our national economy during the first half of the nineteenth century, many of the unions formed survived only for brief periods. However, during the 1850s several national unions were founded. By 1859, the Stonecutters, Hat Finishers, Molders, Machinists, and Locomotive Engineers had emerged as national organiztions. As national labor unions grew in size and number, collective bargaining began to gain momentum.

By 1850, the workday, often from sunrise to sunset early in the century, was shortened to ten hours for most skilled artisans in large cities. Also, wages for the cities' skilled crafts workers increased from $1.25-$1.50 per day in 1820 to $1.50-$2.00 or more by 1860.

While bargaining with employers was considered to be the most important objective of the early labor movement, labor unions took on another function—political activity. They became active in government elections by supporting candidates who were sympathetic to their cause and by getting laws passed that were favorable to labor. One major social gain made by unions was the extension of citizenship rights. In the decades before the Civil War, the right to vote was denied to working people who did not own property; but unions fought and gradually won the battle, giving these people the right to vote.

Post-Civil War

The 15 years following the Civil War comprised an important formative period of the American labor movement. During two cycles of economic recession and revival, 14 new national unions were formed. Union membership rose to 300,000 by 1872, and then dropped to 50,000 by 1878. Three unsuccessful attempts were made to unite the various craft organizations into national labor federations. This period also marked the rise of the eight-hour day movement which was highlighted in 1868 when Congress established an eight-hour workday for federal employees. By 1885, total union membership again reached the 300,000 level.

In 1869, the Noble Order of the Knights of Labor was founded. It consisted of a small local union of Philadelphia garment workers. From an estimated membership of 10,000 in 1879, the Knights of Labor grew rapidly until, in 1886, it had a national membership of more than 700,000. The Knights of Labor's organization was based upon the direct affiliation of local unions with the national association.

After the emergence of the American Federation of Labor (AFL), the Knights steadily lost ground and in 1890 reported only 100,000 members. The Order continued to lose members and finally ceased to be an influential factor in the labor movement, although it continued in existence until 1917.

American Federation of Labor

By 1881, the nucleus of a new organization had taken shape. The American Federation of Labor favored forming a national organization consisting of national trade or craft unions. In the three decades following 1890, it consolidated its position as the principal federation of American unions. By 1920, its membership had grown to more than four million.

The labor movement was, by this time, functioning as a viable organ under the patronage of the federation and accomplishing some of its goals in terms of economic and social improvements. However, it was not free of opposition and setbacks. Nonetheless, as labor unions approached maturity, they were able to obtain the passage of favorable labor legislation. The enactment of the Clayton Anti-Trust Act (1914), the Davis-Bacon Act (1931), the Norris-LaGuardia Act (1932), and the Wagner Act (1935) were hailed by labor as milestone accomplishments. Each afforded organized labor the opportunity to establish itself more firmly as the medium through which workers could collectively express their needs and desires.

As the membership of the labor movement grew, new problems began to emerge. An internal struggle developed in the American Federation of Labor over the question of whether unions should be organized to include all workers in an industry or should they adhere strictly to certain crafts or occupations. In 1934, the AFL realized that new methods would have to be used to organize industrial firms using assembly line set-ups geared for mass production. Thus, out of the 1934 AFL Convention came a resolution which directed the executive council to issue charters to national and international unions in the automotive, cement, aluminum, and other similarly structured industries, as deemed necessary by the council. The resolution also stated that the jurisdictional rights of existing trade unions would be recognized.

Craft unions would continue to organize those industries where skills and work assignments were distinguishable among crafts. However, the industrial or craft organization issue remained unresolved. The problem eventually led to a division in the labor movement which lasted until the merger between the American Federation of Labor and the Congress of Industrial Organizations (CIO) in December, 1955. The Congress of Industrial Organizations was an outgrowth of the committee which had been formed by a few AFL unions to promote the organization of workers on an industry-wide basis rather than along trade or craft lines.

Labor Legislation

During the transition period when the United States shifted from an economy geared to war production to one focusing on peace, there was a certain amount of industrial strife and work stoppage. This created an undesirable situation for organized labor. It was faced with revived, and greatly strengthened, opposition to the Wagner Act—the most significant labor law that had been enacted thus far. Senator Robert Taft and Congressman Fred Hartley sponsored a rewriting of the act which led to the passage of the Labor-Management Relations Act in 1947. The act, often called the Taft-Hartley Act, became law on June 23, 1947, despite organized labor's strong objections and presidential veto. In 1959, the Labor-Management Reporting and Disclosure Act (Landrum-Griffin Act) was enacted. The intent of this act is to eliminate or prevent improper practices on the part of labor organizations, employers, labor relations consultants, or their officers and representatives.

Organized labor continued to expand its ranks through the fifties and sixties, reaching a membership in the neighborhood of 20 million. Following that period, the ranks of organized labor grew to about 25 million, but in the 1980s membership declined.

While the labor movement is still primarily concerned with negotiating with employers and organizing nonunion workers, it is also concerned with public issues. Unions have always encouraged labor's participation in national affairs and have sought to strengthen labor's influence on national policy. Numerous times each year, representatives of the AFL-CIO present labor's views to congressional committees conducting hearings on legislation of interest to the trade union movement. Staff members of the AFL-CIO Department of Legislation and of affiliated unions frequent Capitol Hill to discuss pending bills with individual representatives and senators. Thus, labor unions have become effective in making their voices heard in the nation's legislative councils, as well as at the bargaining table.

In the 1990s an important objective of organized labor will be the promotion of an economic program that will provide American workers the opportunity for jobs and the right to a full-employment economy. Organized labor clearly recognizes that the underutilization of human resources is a terrible waste. The economic viability of our economy is highly dependent on a fully employed work force. Our nation's industries are more productive when they are operating at full employment and plant capacity. Our social welfare programs work more effectively when there is little unemployment. The need for full employment is apparent. Congress recognized this need and enacted the Humphrey-Hawkins Bill for the purpose of achieving full employment.

Organized labor will continue to work for passage of legislation that is beneficial to the workers of our nation. Laws that establish minimum wage levels, protect workers' pension benefits, workplace environments, safety and health, fair trade, and public education are examples of the kinds of legislation organized labor supports. Organized labor has long been the champion for public education and laws that provide for social justice. The American labor movement has always stood for a strong economy and industrial base. Just as in previous decades, organized labor in the 1990s will continue its efforts to fulfill these objectives.

INTERNATIONAL BROTHERHOOD OF ELECTRICAL WORKERS

The International Brotherhood of Electrical Workers (IBEW) is the union which represents most of the organized workers employed in the various occupations discussed in this book. The IBEW was founded in 1891 in St. Louis, Missouri, by a group of electrical workers who were seeking to improve their working conditions. Out of this nucleus has grown an organization which today encompasses a membership of nearly a million, including 1,500 local unions spread across the United States and Canada.

IBEW members are employed in every facet of the electrical industry. Some of the major areas in which they are employed include inside and outside construction, utilities, communications, railroads, electrical manufacturing, and government facilities. Many members are engaged in the crucial work of keeping Americans and Canadians healthy and their countries strong. Some work with radar and on nuclear projects; some do x-ray and other medical work. In addition, thousands of IBEW members are involved in the national space program, handling timing and firing, instrumentation, data acquisition and processing, and communications support services.

The IBEW is structured to provide every member a voice in its operations. Hence, members have the opportunity to express their opinions through the officers they choose to represent them at the local union level. In turn, the International Union, an affiliate of the AFL-CIO, is able to register the views of its entire membership through the federation. This results in an effective and influential body capable of accomplishing many of its objectives.

The objectives of the IBEW can be outlined best by restating them as they appear in its constitution:

> To organize all workers in the entire electrical industry, including all those in public utilities and electrical manufacturing, into local unions; to cultivate feelings of friendship among those of our industry; to settle all disputes between employers and employees by arbitration, if possible; to assist each other in sickness or distress; to secure employment; to reduce the hours of daily labor; to secure adequate pay for our work; to seek a higher and higher standard of living; to seek security for the individual; and by legal and proper means to elevate the moral, intellectual, and social conditions of our members, their families, and dependents, in the interest of a higher standard of citizenship.

NATIONAL ELECTRICAL CONTRACTORS ASSOCIATION

The National Electrical Contractors Association (NECA) is a nationwide trade association which represents the electrical contracting industry. The association provides management and technical service to over 4,800 members, most of whom are small businesspeople. As individuals, they generally cannot afford such a wide range of services themselves, but, by sharing costs with others within the association, member electrical contractors are able to enjoy many benefits at a substantial savings.

Proper management services are very important to the small business operator. Some of the services provided by NECA include representation in labor relations, marketing services, technical training services, business management training, public relations, information services, and a field service which covers the whole range of NECA services. NECA's main purpose is to make this information available to electrical contractors and to motivate them to use it to their advantage (and thus to the advantage of the consumer).

For nearly 90 years, the National Electrical Contractors Association has dedicated itself to a constant effort to improve the quality of electrical service offered to the American public and to industry. NECA and the International Brotherhood of Electrical Workers cosponsor the National Joint Apprenticeship and Training Committee for the Electrical Contracting Industry. This committee sets standards and assists local Joint Apprenticeship and Training Committees, to ensure that, in every locality, an adequate supply of thoroughly trained electrical workers will be available to serve the industry's needs.

In 1920, NECA and the International Brotherhood of Electrical Workers formed a joint committee to deal with the various problems that arise in the field of labor-management relations. This joint committee, consisting of equal employer and union representation, became known as the Council on Industrial Relations (CIR) for the electrical contracting industry. CIR is, in essence, a judicial body rather than a mere arbitration organ. Since its inception, it has functioned in an effective manner, reconciling disputes between the employer and the union, thereby resulting in a virtually strikeless industry.

NATIONAL ELECTRIC SIGN ASSOCIATION

The National Electric Sign Association (NESA) is incorporated as a nonprofit corporation to promote the welfare of the electric sign industry. In 1944, the present National Electric Sign Association—now composed of most of the major electric sign producers, suppliers, and component-part manufacturers—was organized.

The association has continued to gain strength as a guiding force and voice for the industry. Today, NESA has about a thousand members. Groups represented by NESA include custom and national sign companies, sign product manufacturers, sign supply distributors, affiliated state associations, advertisers (sign users), and related organizations.

NESA provides its members with current industrial data such as economical and statistical analyses pertinent to the electric sign industry. The association has a file on job classifications within the sign industry. It also issues a monthly newsletter and publishes a membership directory as well as other special publications.

ELECTRICAL APPARATUS SERVICE ASSOCIATION

Another active trade association in the electrical industry is the Electrical Apparatus Service Association, more commonly known as EASA. EASA is an organization of independent motor repair firms engaged in the repair, maintenance, and sale of electric motors, transformers, generators, controls, and associated equipment. Founded in 1933, EASA is a nonprofit organization with a membership of over 2,700 service centers. Membership is voluntary and open to electrical apparatus service firms that meet membership requirements. The income derived by the association is spent solely for the benefit of its members.

The EASA headquarters staff serves as a research and development team for each member's company—providing information in every conceivable area that might contribute to the further success of the member's business. Each member is kept informed of any trends or activities that affect the electrical repair industry. Members receive new information, technical data, advertising and promotional services, management aids, educational and training programs, and many other services and benefits that have been created specifically to meet their needs in the electrical motor shop and apparatus repair industry.

EDISON ELECTRIC INSTITUTE

The Edison Electric Institute (EEI) is an association of electric light and power companies in the United States. Its affiliate members consist of investor-owned electric utilities. The institute's objectives are advancement in the public service of producing, transmitting, and distributing electricity; promotion of scientific research in such fields; ascertaining and making available to members and to the public data and statistics relating to the electric industry; and aiding operating company members in generating and selling electric energy at the lowest possible price commensurate with safe and adequate service, giving due regard to the interests of consumer, investor, and employee.

The Edison Electric Institute was organized in 1933. Since that time, it has been a strong, continuous stimulant to the advancement of the art of making electricity. Through the years, as new devices and techniques are developed, their benefits are made known to all electric companies and are passed on by the companies to customers throughout the country. EEI is the forum through which this information is exchanged.

INTERNATIONAL ASSOCIATION OF ELECTRICAL INSPECTORS

The International Association of Electrical Inspectors (IAEI) is a nonprofit organization cooperating in the formulation and uniform application of standards for the safe installation and use of electricity. The association has a membership of more than 18,500, representing every branch of the electrical and allied industries. Some of its major objectives are as follows: formulation of standards for the safe installation and use of electrical materials, devices, and appliances; promotion of uniform understanding and application of the National Electrical Code, other codes, and any adopted electrical codes in other countries; promotion of uniform administrative ordinances and inspection methods; collection and dissemination of information relative to the safe use of electricity; representation of electrical inspectors; cooperation with other national and international organizations in the further development of the electrical industry; promotion of cooperation among inspectors, inspection departments, the electrical industry, and the public.

NATIONAL FIRE PROTECTION ASSOCIATION

Since its inception in 1896, the National Fire Protection Association (NFPA) has been a nonprofit, voluntary membership organization. Its membership includes more than 47,000 members worldwide. It is recognized internationally as a clearinghouse for information on fire prevention, fire-fighting procedures, and means of fire protection, and as an authoritative source for fire loss experience.

Membership in NFPA is open to all those interested in promoting the science and improving the methods of fire protection and prevention in all or any of its many facets. Its membership is widely representative of industry, commerce, the fire services, government agencies at all levels, the military forces, architects, engineers, the other professions, hospital and school administrators, and others who have vocational or avocational interests in protection from fires. NFPA's basic objective is to provide humanity with a fire-safe environment, using scientific techniques and education. Its basic function is the preparation of consensus standards and codes relating to fire protection and prevention, with safety the paramount concern.

NFPA has an Electrical Section which provides an opportunity for members interested in electrical safety to become better informed and to contribute to the development of the National Electical Code and other NFPA electrical standards.

The National Electrical Code (NEC) is sponsored by NFPA under the auspices of the American National Standards Institute (ANSI). A primary function of the National Electrical Code Committee is developing periodic revisions of the NEC. The National Fire Protection Association has acted as sponsor of the NEC since 1911. The original code document was developed in 1897, as a result of the united efforts of various insurance, electrical, architectural, and allied interests.

The purpose of the National Electrical Code is the practical safeguarding of persons and of buildings and their contents from hazards arising from the use of electricity for light, heat, power, radio and signalling, and other purposes. The NEC is purely advisory, as far as the NFPA and ANSI are concerned, but is offered for use in law and for regulatory purposes in the interest of life and property protection. The National Fire Protection Association is located at Batterymarch Park, Quincy, Massachusetts 02269.

NATIONAL ELECTRICAL MANUFACTURERS ASSOCIATION

The National Electrical Manufacturers Association (NEMA) is the industry spokesperson on standardization matters for its seven product divisions: building equipment, electronics, industrial equipment, insulating materials, lighting equipment, power equipment, and wire and cable.

NEMA is the nation's largest trade organization for manufacturers of electrical products, and includes 600 member companies located throughout the United States. All are domestic firms, varying in size from comparatively small companies to diversified industrial giants.

Because standards are recognized as essential to industrial progress, NEMA devotes a substantial portion of its budget and a major portion of its staff time to engineering and technical programs. As a result, NEMA today occupies a prominent and highly respected position in both domestic and international standardizations.

The association is a strong advocate of voluntary standards as being necessary to provide sound and safe electrical products to all people in all markets. At the same time, it recognizes that "any standards are good only to the degree that they assist in performing the desired job more efficiently, more economically, and to the greater satisfaction of all concerned."

REGULATORY AND LICENSING AUTHORITIES

Many states, and most of their larger political subdivisions, have adopted regulations and standards governing electrical installations. However, in some instances, these regulations do not apply to certain types of installations. Usually, when either a state or city has an electrical code—normally the National Electrical Code (NEC)—some means of enforcement is also provided.

A large number of states and municipalities attempt to regulate the conformity of electrical work with existing codes by licensing those who are qualified to interpret the pertinent codes. Most states, and a substantial number of cities, require that electrical contractors be licensed, since they are primarily responsible for the electrical work performed by the workers in their employ. A lesser number of states and communities require that journeymen electricians also be licensed. Many states and cities employ electrical inspectors who are given the responsibility of determining whether or not electrical installations meet the requirements of the National Electrical Code and/or other codes which pertain thereto. Some communities employ private inspection agencies.

The National Electrical Code has been widely adopted by state law or city ordinance; it is universally recognized throughout the United States and is the basis for practically all the legislation adopted regarding electrical installations in buildings.

APPENDIX A

SOURCES OF ADDITIONAL INFORMATION

International Brotherhood of Electrical Workers
1125 15th Street N.W.
Washington, DC 20005

National Joint Apprenticeship and Training Committee for the Electrical Industry
16201 Trade Zone Ave.
Suite 105
Upper Marlboro, MD 20772

National Electrical Contractors Association
7315 Wisconsin Ave.
Washington,DC 20014

Edison Electric Institute
1140 Connecticut Ave., N.W.
Washington, DC 20036

Electrical Apparatus Service Association, Inc.
1331 Baur Blvd.
St. Louis, MO 63132

U.S. Department of Labor
  Bureau of Apprenticeship and Training
  200 Constitution Ave.
  Washington, DC 20210

APPENDIX B

BUREAU OF APPRENTICESHIP AND TRAINING

Regional Offices

Location	*States served*	
Room 510 John F. Kennedy Fed. Bldg. Government Center Boston, MA 02203	Connecticut Maine Vermont	Massachusetts New Hampshire Rhode Island
Room 602 201 Varick St. New York, NY 10036	New Jerse New York	Puerto Rico Virgin Islands
Gateway Bldg. 3535 Market St. Philadelphia, PA 19104	Delaware Maryland Virginia	Pennsylvania West Virginia
Room 418 1371 Peachtree St., N.E. Atlanta, GA 30309	Alabama Florida Georgia Kentucky	Mississippi North Carolina South Carolina Tennessee

Room 758 Federal Building 230 South Dearborn St. Chicago, IL 60604	Illinois Indiana Michigan	Minnesota Ohio Wisconsin
Room 502, Federal Building 525 Griffin St. Dallas, TX 75202	Arkansas Louisiana New Mexico	Oklahoma Texas
Room 1100, Fed. Office Bldg. 911 Walnut St. Kansas City, MO 64106	Iowa Kansas	Missouri Nebraska
Room 476 New Custom House 721 19th St. Denver, CO 80202	Colorado Montana Utah	North Dakota South Dakota Wyoming
Room 715 71 Stevenson St. San Francisco, CA 94105	Arizona California	Hawaii Nevada
Federal Office Building Room 8018 909 First Ave. Seattle, WA 98174	Alaska Idaho	Oregon Washington

Federal Offices

ALABAMA
Suite 102
Berry Building
2017 - 2nd Avenue, N.
Birmingham 35203

ALASKA
Room C-528
Federal Building and Courthouse, Box 37
701 C Street
Anchorage 99513

ARIZONA
Suite 302
3221 North 16th Street
Phoenix 85016

ARKANSAS
Room 3014
Federal Building
700 West Capitol Street
Little Rock 72201

CALIFORNIA
Room 350
211 Main Street
San Francisco 94105

COLORADO
Room 480
U.S. Custom House
721 - 19th Street
Denver 80202

CONNECTICUT
Room 367
Federal Building
135 High Street
Hartford 06103

DELAWARE
Lock Box 36
Federal Building
844 King Street
Wilmington 19801

FLORIDA
Room 1049
City Centre Building
227 North Bronough Street
Tallahassee 32301

GEORGIA
Room 418
1371 Peachtree Street, N.E.
Atlanta 30367

HAWAII
Room 5113
P.O. Box 50203
300 Ala Moana Boulevard
Honolulu 96850

IDAHO
Room 493
P.O. Box 006
550 West Fort Street
Boise 83724

ILLINOIS
Room 758
230 S. Dearborn Street
Chicago 60604

INDIANA
Room 414
Federal Building and U.S. Courthouse
46 East Ohio Street
Indianapolis 46204

IOWA
Room 637
Federal Building
210 Walnut Street
Des Moines 50309

KANSAS
Room 235
Federal Building
444 S.E. Quincy Street
Topeka 66683

KENTUCKY
Room 187-J
Federal Building
600 Federal Place
Louisville 40202

LOUISIANA
Room 1323
U.S. Postal Building
701 Loyola Street
New Orleans 70113

MAINE
Room 408-D
Federal Building
P.O. Box 917
68 Sewall Street
Augusta 04330

MARYLAND
Room 1028
Charles Center - Federal Building
31 Hopkins Plaza
Baltimore 21201

MASSACHUSETTS
Room 1703-B
JFK Federal Building
Government Center
Boston 02203

MICHIGAN
Room 657
Federal Building
231 W. Lafayette Avenue
Detroit 48226

MINNESOTA
Room 134
Federal Building and U.S. Courthouse
316 Robert Street
St. Paul 55101

MISSISSIPPI
Suite 1010
Federal Building
100 West Capitol Street
Jackson 39269

MISSOURI
Room 547
210 North Tucker
St. Louis 63101

MONTANA
Room 394 - Drawer #10055
Federal Office Building
301 South Park Avenue
Helena 59626-0055

NEBRASKA
Room 700
106 South 15th Street
Omaha 68102

NEVADA
Room 311
Post Office Building
P.O. Box 1987
301 East Stewart Avenue
Las Vegas 89101

NEW HAMPSHIRE
Room 311
Federal Building
55 Pleasant Street
Concord 03301

NEW JERSEY
Room 339
Military Park Building
60 Park Place
Newark 07102

NEW MEXICO
Suite 16
320 Central Avenue, S.W.
Albuquerque 87102

NEW YORK
Room 810
Federal Building
North Pearl & Clinton Avenues
Albany 12201

NORTH CAROLINA
Room 376
Federal Building
310 New Bern Avenue
Raleigh 27601

NORTH DAKOTA
Room 344
New Federal Building
653 - 2nd Avenue, N.
Fargo 58102

OHIO
Room 605
200 North High Street
Columbus 43215

OKLAHOMA
Room 526
Alfred P. Murrah Federal Building
200 N.W. Fifth Street
Oklahoma City 73102

OREGON
Room 526
Federal Building
1220 S.W. 3rd Avenue
Portland 97204

PENNSYLVANIA
Room 773
Federal Building
228 Walnut Street
Harrisburg 17108

RHODE ISLAND
100 Hartford Avenue
Federal Building
Providence 02909

SOUTH CAROLINA
Room 838
Strom Thurmond Federal Building
1835 Assembly Street
Columbia 29201

SOUTH DAKOTA
Room 403
Courthouse Plaza
300 N. Dakota Avenue
Sioux Falls 57102

TENNESSEE
Suite 101-A
460 Metroplex Drive
Nashville 37211

TEXAS
Room 2102
VA Building
2320 LaBranch Street
Houston 77004

UTAH
Room 1051
1745 West 1700 South
Salt Lake City 84104

VERMONT
Suite 103
Burlington Square
96 College Street
Burlington 05401

VIRGINIA
Room 10-020
400 North 8th Street
Richmond 23240

WASHINGTON
Room B-104
Federal Office Building
909 First Avenue
Seattle 98174

WEST VIRGINIA
Room 310
550 Eagan Street
Charleston 25301

WISCONSIN
Room 303
Federal Center
212 East Washington Avenue
Madison 53703

WYOMING
Room 8017
J.C. O'Mahoney Federal Center
P.O. Box 1126
2120 Capitol Avenue
Cheyenne 82001

APPENDIX C

HELPFUL PUBLICATIONS

National Apprenticeship Program: U.S. Department of Labor/Employment and Training Administration, Bureau of Apprenticeship Training. 200 Constitution Avenue, N.W., Washington, DC 20210

Apprenticeship Information: U.S. Department of Labor/Employment and Training Administration. 200 Constitution Avenue, N.W., Washington, DC 20210

Apprenticeship—Past and Present: U.S. Department of Labor/Employment and Training Adminstration, Bureau of Apprenticeship and Training. 200 Constitution Avenue, N.W., Washington, DC 20210

Electrical Apprenticeship: National Joint Apprenticeship and Training Committee for the Electrical Industry. 16201 Trade Zone Avenue, Suite 105, Upper Marlboro, MD 20772

Federal Benefits for Veterans and Dependents: Fact Sheet 1S-1, Department of Veterans Affairs, Washington, DC 20420

Directory of Post Secondary Schools: With Occupational Programs: U.S. Department of Education, National Center for Education Statistics, Washington, DC 20402

Occupational Outlook Handbook: Bulletin 2300, U.S. Department of Labor, Bureau of Labor Statistics. (This Handbook can usually be found in the office of high school guidance counselors and public libraries.)

APPENDIX D
GLOSSARY

Agreement, collective bargaining. A written agreement (contract) arrived at as the result of negotiation between an employer or a group of employers and a union. It sets the conditions of employment—wages, hours, fringe benefits—and the procedure to be used in settling disputes that may arise during the term of the contract. The term of a contract may be for one, two, or three years.

Apprentice. Usually a young person who enters into agreement to learn a skilled trade and to achieve a journeyman status through supervised training and experience, usually for a specified period of time. Practical training is supplemented by related technical off-the-job instruction.

Building trades. The skilled trades in the building industry. These include: electricians, carpenters, plumbers, painters, plasterers, bricklayers, and stonemasons. These crafts generally have well-developed apprenticeship programs.

Collective bargaining. A method of determining conditions of employment by negotiation between representatives of the employer and union representatives of the employees. The results of collective bargaining are set forth in a collective bargaining agreement.

Compensation. A concept sometimes used to encompass the entire range of wages and benefits, both current and deferred, which workers receive out of their employment.

Craftsperson. An artisan. A person skilled in the mechanics of his or her craft.

Earnings. The total amount of remuneration received by a worker for a given period as compensation for work performed or services rendered, including incentive pay, premium pay for overtime, shift differentials, bonuses, and commissions.

Fringe benefits. Nonwage benefits and payments received by or credited to workers in addition to wages. Examples are: supplemental unemployment benefits, pensions, travel pay, vacation and holiday pay, and health insurance.

Journeyman. A skilled worker, generally having mastered his or her trade by serving an apprenticeship.

Median wage. That wage rate which occupies the middle position in an array of wage rates. To illustrate the distinction between MEDIAN and AVERAGE, suppose five persons have wage rates respectively of $8, $9, $10, $13, and $15 an hour. The average wage rate is $11 per hour, while the median wage rate is $10 per hour.

Overtime. Work performed in excess of basic workday or workweek, as defined by law, collective bargaining agreement, or company policy. Often applied to work performed on Saturdays, Sundays, and holidays at premium rates of pay.

Service calls. A term generally applied to the work performed by a serviceperson or craftsperson at the customer's premises.

Shift work. The term applied to the daily working schedule of a firm or its employess. Day shift—usually the daylight hours; evening shift—work schedule ending at or near midnight; night (graveyard) shift—work schedule starting at or near midnight.

Troubleshoot. The term applied to locating and/or diagnosing faults in electrical circuits, appliances, machinery, and equipment.

APPENDIX E

STATE APPRENTICESHIP AGENCIES

Arizona

Apprenticeship Services
  Department of Economic Security
  438 West Adams Street
  Phoenix, AZ 85003

California

Division of Apprenticeship Stds.
  Dept. of Industrial Relations
  P.O. Box 603
  525 Golden Gate Avenue
  San Francisco, CA 94101

Colorado

Colorado Apprenticeship Council
  State Centennial Building
  Room 314
  1313 Sherman Street
  Denver, CO 80203

Connecticut

Apprenticeship Training Division
   Labor Department
   200 Folly Brook Boulevard
   Wethersfield, CT 06109

Delaware

Apprenticeship and Training Council
   Department of Labor
   Division of Employment and Training
   6th Floor—State Office Building
   820 North French Street
   Wilmington, DE 19801

District of Columbia

Director of Apprenticeship
   500 C Street, N.W.
   Suite 241
   Washington, DC 20001

Florida

Bureau of Apprenticeship
   Division of Labor, Employment and Training
   1320 Executive Center Drive
   Tallahassee, FL 32301

Hawaii

Apprenticeship Division
   Dept. of Labor and Industrial Relations
   830 Punch Bowl Street
   Honolulu, HI 96813

Kansas

Apprenticeship Section
  Division of Labor-Management Relations and Employment
    Standards
  Kansas Department of Human Resources
  610 West 10th—2nd Floor
  Topeka, KS 66612

Kentucky

Apprenticeship and Training
  Department of Labor
  Division of Labor Standards
  620 S. Third Street
  Louisville, KY 40202

Louisiana

Director of Apprenticeship
  Department of Labor
  5360 Florida Boulevard
  Baton Rouge, LA 70806

Maine

Director of Apprenticeship
  Bureau of Labor Standards
  State House Station #45
  Augusta, ME 04333

Maryland

Apprenticeship and Training
  Department of Employment and Training
  1100 N. Eutaw Street
  Baltimore, MD 21202

Massachusetts

Division of Apprentice Training
  Department of Labor and Industries
  Leverett Saltonstall Building
  100 Cambridge Street
  Boston, MA 02202

Minnesota

Division of Apprenticeship
  Department of Labor and Industry
  Space Center Bldg.—4th Floor
  443 Lafayette Road
  St. Paul, MN 55101

Montana

Appenticeship and Training
  Montana Department of Labor and Industry
  P.O. Box 1728
  Jackson Street Entrance
  Helena, MT 59620

Nevada

Nevada Apprenticeship Council
  Department of Labor
  505 East King Street—Room 601
  Carson City, NV 89710

New Hampshire

Commissioner of Labor
  New Hampshire Apprenticeship Council
  Department of Labor
  19 Pillsbury Street
  Concord, NH 03301

New Mexico

Director/Apprenticeship Section
  New Mexico Department of Labor
  501 Mountain Road, N.E.
  Albuquerque, NM 87102

New York

Director Employability Development
  Department of Labor
  The Campus Building, #12
  Albany, NY 12240

North Carolina

Director
  North Carolina Department of Labor
  Memorial Building
  214 W. Jones Street
  Raleigh, NC 27603

Ohio

Ohio State Apprenticeship Council
  Department of Industrial Relations
  2323 West Fifth Avenue
  Columbus, OH 43216

Oregon

Apprenticeship and Training Division
  State Office Building, Room 405
  1400 S.W. Fifth Avenue
  Portland, OR 97201

Pennsylvania

Apprenticeship and Training
   7th and Forester Streets
   Department of Labor and Industry
   Labor and Industry Building
   Harrisburg, PA 17120

Puerto Rico

Director, Incentive to the Private Sector Program
   Right to Employment Administration
   P.O. Box 4452
   San Juan, PR 00936

Rhode Island

Rhode Island Apprenticeship Council
   Department of Labor
   200 Elmwood Avenue
   Providence, RI 02907

Utah

Utah Apprenticeship Council
   28 East 2100 South
   Chapman Plaza Building, Suite 104
   Salt Lake City, UT 84115

Vermont

Director
   Vermont Apprenticeship Council
   Department of Labor and Industry
   120 State Street
   Montpelier, VT 05602

Virginia

Division of Apprenticeship Training
   Department of Labor and Industry
   205 North 4th—4th and Grace Streets
   P.O. Box 12064
   Richmond, VA 23241

Virgin Islands

Division of Apprenticeship and Training
   Department of Labor
   P.O. Box 890
   Christiansted, Saint Croix
   VI 00820

Washington

Director, State of Washington
   Department of Labor and Industries
   ESAC Division
   925 Plum Street
   Olympia, WA 98504

Wisconsin

Bureau Director
   Department of Industry, Labor and Human Relations
   P.O. Box 7972
   Madison, WI 53707

621.3
Wo

C. 1

Wood, Robert

Opportunities in electrical trades

DATE DUE

South Hunterdon Reg. Library
301 Mt. Airy-Harbourton Rd.
Lambertville, NJ 08530

4255 West Touhy Avenue
Lincolnwood, Illinois 60646-1975 U.S.A.